# NORMATIVE ETHICS

Other books from Automatic Press ◆ $\frac{V}{I}$P

Formal Philosophy
edited by Vincent F. Hendricks & John Symons
November 2005

Thought$_2$Talk: A Crash Course in Reflection and Expression
by Vincent F. Hendricks
September 2006

Masses of Formal Philosophy
edited by Vincent F. Hendricks & John Symons
October 2006

Political Questions: 5 Questions for Political Philosophers
edited by Morten Ebbe Juul Nielsen
December 2006

Philosophy of Technology: 5 Questions
edited by Jan-Kyrre Berg Olsen & Evan Selinger
February 2007

Game Theory: 5 Questions
edited by Vincent F. Hendricks & Pelle Guldborg Hansen
April 2007

Legal Philosophy: 5 Questions
edited by Morten Ebbe Juul Nielsen
October 2007

Philosophy of Mathematics: 5 Questions
edited by Vincent F. Hendricks & Hannes Leitgeb
December 2007

Philosophy of Physics: 5 Questions
edited by Juan Ferret & John Symons
March 2008

Probability and Statistics: 5 Questions
edited by Alan Hájek & Vincent F. Hendricks
March 2008

# NORMATIVE ETHICS
## 5 QUESTIONS

edited by

Thomas S. Petersen

Jesper Ryberg

Automatic Press ♦ $\frac{\text{V}}{\text{I}}$P

Automatic Press ◆ $\frac{\text{V}}{\text{I}}$P

Information on this title: www.normativeethics.com

© Automatic Press / VIP 2007

First published 2007

Printed in the United States of America
and the United Kingdom

ISBN-13    978-87-92130-00-6    paperback

Typeset in LaTeX2$_\varepsilon$
Graphic design by Vincent F. Hendricks

# Contents

ii

# Preface

How ought we to live our lives? When is an act morally right or wrong? Or more specifically, what ought we, for instance, do about the horrible fact that 40.000 children die every day because of starvation or malnutrition? Is war always morally wrong? Is abortion? Questions like these are discussed systematically within *normative ethics*. Normative ethics, as a philosophical discipline, tries to clarify and critically discuss moral concepts and principles that can be used to answer such questions.

*Normative Ethics: 5 Questions* is a collection of short interviews with 18 of the most influential philosophers in normative ethics. Starting with the biographical question of why they were initially drawn to their field, the interview proceeds with questions about their views on normative ethics, its aim, scope, prospect and future direction and how their work fits in these regards. The collection, which, to our knowledge and surprise, is completely unprecedented, is both a "popular" way of conveying the importance of normative ethics and gives insight to students, teachers and researchers into how leading scholars view their own field. More precisely the 5 questions we posed were:

1. Why were you initially drawn to normative ethics?

2. What example(s) from your work (or the work of others) illustrates the role that normative ethics ought to play in moral philosophy?

3. How do studies within scientific disciplines contribute to the development of normative ethics?

4. What do you consider the most neglected topics and/or contributions in normative ethics?

5. What are the most important problems in normative ethics and what are the prospects for progress?

We posed these meta-philosophical questions because of our curiosity and because they are very seldom answered in the philosophical literature on normative ethics. That the latter is the case should come as no surprise. In philosophical journals or books our questions are either conceived of as irrelevant (e.g. question 1) or the answers are presupposed in the mere publishing of a paper in e.g. a journal (e.g. question 2). When people publish papers in journals it is usually because they believe that the paper exemplifies the role normative ought to play in moral philosophy. However, curiosity and rareness may not seem to be sufficient reasons for engaging in such a project.

There is a multiplicity of reasons, besides the two already mentioned, why we chose to ask exactly these meta-philosophical questions are many. Let us mention a few of these. We believe that it would be interesting and inspiring to know what has motivated some of the most influential moral philosophers of our time to spend most of their research hours on doing normative ethics. Besides all the arguments, distinctions and theories that philosophers present in their work we wanted to know something about the persons behind the work. Although normative ethics can flourish without such biographical knowledge, it is our hope that biographical data and stories can make us more interested in reading normative ethics and maybe help us to better understand some of the philosophical texts. Furthermore, by giving these moral philosophers a chance to present what they believe are their (or others) most valuable contributions to normative ethics, the reader may become aware that he or she has overlooked important contributions by some of these thinkers. And as many moral philosophers are engaged with moral problems that arises because of scientific knowledge (e.g. within biotechnology, war-technology or psychology) and because some philosophers or scientists believe that moral philosophy can be reduced to science, we thought it would be illuminating to ask a question about the interplay between science and moral philosophy.

By reading these interviews you will learn how a communist mother or how sadistic teachers made some of these scholars interested in normative ethics. You will learn how one philosopher woke one morning and discovered that he was a utilitarian or another who where not initially drawn to normative ethics found herself publishing books on applied ethics. Besides these biographical stories which are written with extreme honesty and sometimes in a very humorous way, you will get an idea of what these thinkers believe are the central and most important topics within normative ethics and what the prospect for progress concerning these topics amounts to. As some of the philosophers says in this book, philosophers should not only be concerned with meta-ethics (e.g. "Whether moral judgements have truth value or not?") or metaphysics (e.g. "How do I know that other people exist?") when we live in a world where millions are killed by war, famine or diseases.

We have allowed the responders the freedom to express there answer to these meta-philosophical questions in any which way they would prefer. As a consequence of this, some of the answers are as long as 6000 words whereas others are about 2000 words. Most have preferred to answer all the questions and a few have chosen to answer some of them. Some have chosen to write an essay where there is no sharp division between the 5 questions, while most have answered the questions one by one.

Thomas S. Petersen & Jesper Ryberg
Roskilde University, Denmark
November 2007

# Acknowledgements

We would like to thank Automatic Press ♦ ⌄⌐P and its founder V.J. Menshy for having invited us to edit this book. Special thanks to Vincent for his idea about the interview book concept and for his ongoing and inspiring interest in our specific interview book. Thanks to Jakob v.H. Holtermann for valuable comments to the preface. Thanks also to Claus Festersen for translating the entire document into LaTeX. Finally, we would of course like to express warm thanks to all the moral philosophers who have answered the 5 questions and for their kind cooperation, encouragement and last but not least—patience. The royalties from this publication go to UNICEF.

Thomas S. Petersen & Jesper Ryberg
Roskilde University, Denmark
November 2007

# 1

# Elizabeth S. Anderson

Arthur Thurnau and John Rawls Collegiate
Professor of Philosophy and Women's Studies

University of Michigan, Ann Arbor, USA

---

**Why were you initially drawn to normative ethics?**

My father, who earned a Ph.D. in engineering, missed the liberal
arts education he never got and decided when I was a teenager to
study the liberal arts by reading philosophical texts with me. We
read Plato's *Republic* and John Stuart Mill's *On Liberty* together.
Issues in moral and political philosophy were constant topics of
conversation at the family dinner table. I drank from the goblet of
libertarianism and free market economics. As an undergraduate
at Swarthmore College, I was initially drawn to economics, and
within philosophy, to logic, the philosophy of mathematics, and
the philosophy of science. However, the study of philosophy led
me to question some of the assumptions of economics, which in
turn led me to question the political prescriptions of free-market
economics. I discovered a taint in the liquid from that goblet. Economics didn't provide me with the tools to analyze the contam-
inant, but moral and political philosophy did. My determination
to get to the bottom of the problems in my libertarian philosophy led me to specialize in ethics in graduate school. I wrote
my dissertation on the ethical limitations of markets, focusing on
controversial cases of commodification, such as contract parenting
(surrogate motherhood), prostitution, and cost-benefit analysis.
The dissertation was not a work purely in political philosophy,
however. Normative ethics laid its foundation, in the form of a
pluralistic theory of value.

**What example(s) from your work (or the work of others) illustrates the role that normative ethics ought to play in moral philosophy?**

Normative ethics grows out of reflection on problems, puzzles, and confusions in our moral lives. Ethics as lived is a seat-of-the-pants affair: we draw evaluative distinctions and improvise rules of conduct on the fly, trying to make sense of and solve local problems. Then we generalize, hoping the solutions we devised for one context will work in another. This practice gives rise to conflicting recommendations in new cases, where more than one principle or distinction we have previously lived by seems to apply. Confrontation with such conflicts sends us in two directions—toward so-called "applied ethics," and toward normative ethics or moral theory, which tries to sort out the confusions, tensions, and contradictions in the normative assumptions of our conduct by means of higher-order normative reflection on our evaluative distinctions and norms. The hope of normative theory is not simply that we can systematize our evaluative intuitions, but that we can offer an account of why they present themselves to us in characteristic patterns, and link that account back to the roles our normative distinctions and judgments play in our actual lives. The hope is that, by getting clearer on what deeper normative notions our off-the-cuff normative judgments (our "intuitions") are tracking, on what could make sense of them in a way we can reflectively endorse, we can lead our lives with greater self-understanding. We will live our lives more meaningfully when we act with a deeper knowledge and appreciation of the normative points of our action. We will also, we hope, be able to resolve the normative conflicts that arise in our lives.

This idea, that normative ethics is inherently connected to the project of elucidating and criticizing the self-understandings implicit in our practices, inspired my book, *Value in Ethics and Economics.*[1] There I highlight an alternative to the monistic "weighing" metaphor that we sometimes use in deliberation between disparate options. The metaphor invites us to think of different goods as all measured on a common scale, differing only in quantity, not quality. The starting point of my alternative, pluralistic theory of value is that we don't simply value different goods *more* or *less*; we value them in fundamentally different *ways*. Love, honor, respect, and esteem are distinct evaluative attitudes we have toward per-

---

[1] Cambridge, Mass. Harvard University Press, 1993.

sons (and some animals). Toward some natural phenomena and artifacts we respond with wonder, fascination, and awe; others we cherish; still others we relish. Different goods merit distinct favorable attitudes. The function of value judgments is to guide these evaluative responses—to guide feelings and attitudes, not just actions and choices. Merited feelings and attitudes, in turn, generate reasons for actions that express these attitudes.

Once we appreciate this plurality of goods in terms of the plurality of our evaluative attitudes, we can arrive at better ways of dealing with conflict cases by considering how well or poorly various principles for treating the good express the kind of valuation it merits. In my book I criticized contract parenting for commodifying mothers and children, which I argued is inconsistent with the respect and love due respectively to them. Since then, I have elaborated on the plurality of values to help us understand conflict cases in environmental ethics.[2] For instance, environmental ethicists advocate eradicating the ecologically destructive rabbit from the Australian outback, whereas animal rights and animal welfare theorists object to such action. Animal welfare theorists may allow some animal experimentation, if the medical treatments thereby validated relieve more human suffering than the experiments caused, whereas animal rights theorists object to experiments that inflict any suffering on animals. Here we are torn by distinct evaluative attitudes—on the environmental ethicist's side, *esteem* for and *appreciation* of the wonders of nature, considered as a complex system of organisms; on the animal rights theorist's side, *respect* for the claims individual animals implicitly make on us to heed their interests; on the animal welfare theorist's side, *sympathy* for all sentient beings. In the first conflict case, I favor the values on the environmental ethicist's side over those on the animal rights theorist's side; and in the second, the values on the animal welfarist's side. But each point of view can claim cases in which its distinctive considerations rightly prevail. We shouldn't always expect normative ethics to give us definitive answers in conflict cases. Its role is not to provide a decision procedure or criterion of right action that works in every case, but rather to highlight different types of normative considerations that might otherwise be neglected, and help us see what roles they should

<hr>

[2] "Animal Rights and the Values of Nonhuman Life," in Martha Nussbaum and Cass Sunstein, eds., *Animal Rights: Current Debates and New Directions* (Oxford: Oxford University Press, 2002).

play in our lives.[3]

I carry out the theme of drawing lessons in normative ethics from critical reflection on one's life and the self-understandings with which one leads it in my work on John Stuart Mill.[4] There I defend the pragmatist point that normative judgments are testable in experience just as scientific hypotheses are. We test our value judgments by putting them into practice, by acting in accordance with them and seeing whether doing so solves the problem we wanted them to solve, or gives rise to worse problems in its wake. John Stuart Mill tested Bentham's monistic hedonism (the doctrine that the best life consists in maximizing pleasure, conceived as a quantifiable feeling that does not differ in quality) in his life and found it wanting. It led him into a depression from which his philosophy offered no means of recovery. By contrast, Mill found relief from depression and genuine happiness in living his life based on an appreciation of pleasures as qualitatively distinct and incommensurable. This experience offered empirical vindication of Mill's pluralistic theory of value over Bentham's. I have also elaborated on this idea of testing value judgments in experience in my work on pragmatism and Dewey's ethics and also in my work vindicating various uses of normative judgments in social scientific research.[5]

My pragmatist approach contrasts with two other methods that are common in normative ethics. One starts with intuitions about particular cases and tries to systematize them into principles of normative ethics. Work on the classic trolley-car dilemmas falls under this category. Some of it is very sophisticated.[6] But pure intuition-mining has little to say about the sources or structure

---

[3] I thank Allan Wood for this point.

[4] "John Stuart Mill and Experiments in Living," *Ethics* 102 (1991): 4-26; "John Stuart Mill on Democracy as Sentimental Education," in Amelie Rorty, ed., *Philosophers on Education* (Chicago: University of Chicago Press, 1998), pp. 333-352.

[5] "Pragmatism, Science, and Moral Inquiry," in Richard Fox and Robert Westbrook, eds. *In Face of the Facts: Moral Inquiry in American Scholarship* (Cambridge: Woodrow Wilson Center Press/Cambridge University Press, 1998), pp. 10-39; "Dewey's Moral Philosophy," *Stanford Encyclopedia of Philosophy* (Spring 2005 Edition), Edward N. Zalta (ed.), URL = <http://plato.stanford.edu/archives/spr2005/entries/dewey-moral/>, 2005; "Uses of Value Judgments in Feminist Social Science: A Case Study of Research on Divorce," *Hypatia* 19 (2004): 1-24.

[6] See, for example, Frances Kamm, *Morality, Mortality* (New York: Oxford University Press, 1993).

of our intuitions, and hence provides limited insight into why our intuitions fall into characteristic patterns. Moreover, it is all armchair normative ethics. It omits what we can learn from considering how our lives go when we actually adopt particular principles of action and conceptions of value. The implications of adopting such principles and conceptions go beyond what they tell us to do in particular cases. They entail certain more general attitudes toward the people in our lives—attitudes that may themselves bear negatively or positively on the quality of our relations to them, over and above how we treat them in life or death decision-making.

The other method starts with universal, abstract normative principles, such as utilitarianism or Kant's categorical imperative, and tries to work out the implications of such principles for particular cases. This is what is properly called "applied ethics"—normative ethical principles as applied to cases. I don't favor this method, either. It assumes too quickly that some very simple principles fully encapsulate all the normatively relevant features of the case. But normative ethics can't be reduced to such simple ideas as promoting welfare or respecting rational agency. Values are too disparate and plural to be encapsulated in one or even a few general principles.

That's why I prefer to conduct normative ethics at a "middle" level: close enough to cases to draw a sense of the rich variety of normatively relevant considerations that shape ethical life, but far enough away to get some perspective on the underlying structure and sources of our moral intuitions. Some kinds of systematization in normative ethics are possible and desirable, because of the light they cast on our self-understandings. We can't rest with a mass of particularistic moral intuitions or with a grab-bag of nearly untheorized reasons. We need to grasp how different types of reasons have special bearing on different types of emotional and attitudinal – that is, evaluative – responses. Normative ethics and moral psychology need to work more closely together in pragmatic normative inquiry.

## How do studies within scientific disciplines contribute to the development of normative ethics?

If judgments in normative ethics can be tested in experience, then the social sciences play a critical role in helping us determine which judgments we should keep and which we should revise—that is, by which judgments we should actually guide our lives. Experience

# 6    1. Elizabeth S. Anderson

must be considered in a broad sense. This means not just personal
experience – although that is vitally important – but the experiences of people living together under common rules of conduct
and shared, or conflicting, conceptions of value. For the social sciences to be helpful in these ways, however, requires that empirical
research methodologies be shaped by our normative concerns. For
example, if we want to determine whether liberalized divorce laws
are good or bad, we need to adopt research methodologies attuned
to values worth caring about.[7] Similarly, if we want to determine
how to assess rival moral economies of credit and debt, we need to
investigate what life was like under these rival systems, and what
values were realized for people under them.[8]

To carry out this pragmatist project, we need to revise crude
accounts of scientific methodology that suppose that science must
be "value neutral." When the question we ask empirical inquiry to
help us answer is itself value-laden, then so must be the concepts,
measuring tools, and methods of that inquiry. The problem is to
figure out how inquiry can be value-laden without being rigged in
advance to favor a conclusion predetermined by the value judgments we are already inclined to hold. This problem is solved by
carefully dividing the labor between normative judgments and evidence, so that judgments don't usurp the roles properly assigned
to evidence in assessing the empirical support for rival hypotheses.
Most of my work in feminist epistemology lies at this methodological intersection of normative and empirical inquiry.[9]

Normative ethics and the social sciences need to work in tan-
dem, with each side being open to revision in light of discoveries
jointly made by both. This should not be surprising, if we conceive
of normative ethics as from the start wrapped up with an elucidation of our self-understandings. For the social sciences are also
exercises in self-understanding. When we adopt certain social scientific theories as our own self-understanding, this has normative

---

[7] See my "Uses of Value Judgments in Feminist Social Science: A Case
Study of Research on Divorce," *Hypatia* 19 (2004): 1-24.

[8] See my "Ethical Assumptions of Economic Theory: Some Lessons from
the History of Credit and Bankruptcy," *Ethical Theory and Moral Practice*
7 (2004): 347-360.

[9] See also "Knowledge, Human Interests, and Objectivity in Feminist Epistemology," *Philosophical Topics* 23 (1995): 27–58; "Situated Knowledge and
the Interplay of Value Judgments and Evidence in Scientific Inquiry," in P.
Gärdenfors, J. Wolenski and K. Kijania-Placek, eds. *In the Scope of Logic,
Methodology and Philosophy of Science*, vol. 2 (Kluwer, 2002), pp. 497–517.

consequences. Charles Taylor has long written about the implications of this fact.[10] Ian Hacking more recently has taken up
this task.[11] The adoption of a certain social scientific picture of
ourselves as who we really are – for instance, a conception of ourselves as rational, self-interested bargainers – can have untoward
consequences, making our practices go less well than before. Social
psychologists and behavioral economists have had a field day with
this point in recent years. Even the seemingly most self-interested
interactions are often governed by social norms of cooperation.
The workplace, for example, would break down if everyone were
narrowly self-interested. More importantly, it breaks down if most
people *think* that most others are narrowly self-interested. That
thought, once it becomes an object of common belief, is a self-
fulfilling prophecy that undermines the conditions of trust that enable cooperation.[12] Recognition of problems of self-understanding
like this led me away from pure theoretical economics to a mode
of doing philosophy that engages economics and the other social
sciences, using their findings in a critical way.

**What do you consider the most neglected topics and/or
contributions in normative ethics?**

Most work in normative ethics is part of ideal theory. We ask
what principles of conduct should guide us, on the assumption
that everyone, or nearly everyone, will or is able to follow those
principles. Moral philosophers rarely ask whether or under what
conditions humans can reasonably be expected to follow those
principles, or can be trusted to do so. It is time that we start
asking these questions in a more serious and systematic way. We
have a lot of evidence from history, anthropology, and the social
sciences, especially cognitive and social psychology, of our considerable moral frailties and cognitive deficiencies. We need to
investigate more thoroughly where our cognitive and behavioral

---

[10] See his *Philosophy and the Human Sciences* (Cambridge: Cambridge University Press, 1985).

[11] Ian Hacking, "Making up People," in *Reconstructing Individualism: Autonomy, Individuality and the Self in Western Thought*, ed. Thomas Heller,
Morton Sosna, and David Wellbery (Stanford, CA: Stanford University Press,
1986), pp. 222-236.

[12] For a review of the literature, see my "Beyond *Homo Economicus*: New
Developments in Theories of Social Norms," *Philosophy and Public Affairs*
29 (2000): 170-200.

defects lie, so that we can devise rules sensitive to them. Normative ethics needs to be re-oriented toward non-ideal theory. Non-ideal theory should work hand-in-hand with a more naturalized moral epistemology–with empirical inquiry into how people reason out moral problems, revise their conceptions of the good, and arrive at new ideals.

What might we be expected to learn from non-ideal theory? Much of my work in normative ethics concerns inequality, especially along lines of race, class, and gender.[13] The institution of systematic inequality along salient social divisions like this tends to distort cognition in predictable ways. Psychologists have established that people develop stereotypes for judging members of different social groups. These stereotypes tend to rationalize the unequal social positions into which members of different races, classes, and genders tend to be placed. They stigmatize those placed in disadvantaged positions. These stereotypes operate unconsciously, behind our backs, in ways that escape deliberate control. We may consciously endorse one belief, but act in accordance with a contrary belief without our being aware of this fact.

This feature of our psychology has implications for the rules we should adopt to guide our lives. In ideal theory, there is a case to be made for the thought that there would be no races in the best society, and hence no reason to ever discriminate according to race. But in non-ideal theory, a norm forbidding all race-based preferences makes no sense. As Aristotle pointed out, when the wind is blowing across the target, you need to aim off-center, against the wind, in order to have any hope of hitting the bulls-eye. Psychologists can help us improve our aim. For instance, we know that racial stigma is reduced, and racial stereotypes are less salient, when people from different racial groups have experience working and cooperating together as equals. Institutionally supported racial integration reduces prejudice and stereotyping.[14] It turns out, however, that under current conditions many institutions cannot be integrated without taking deliberate action—"affirmative

---

[13] "What is the Point of Equality?," *Ethics* 109 (1999): 287-337; "Recent Thinking about Sexual Harassment: A Review Essay," *Philosophy and Public Affairs* 34 (2006): 284-311.

[14] This is the famous "contact" hypothesis originally advanced by Gordon Allport, *The Nature of Prejudice.* (Cambridge, Mass.: Addison-Wesley, 1954) and repeatedly confirmed. For a survey, see Gaertner, S. L., and J. F. Dovidio. 2000. *Reducing Intergroup Bias: The Common Ingroup Identity Model.* Philadelphia: Psychology Press.

action"—to include members of racial groups that are systematically disadvantaged. Non-ideal theory tells us that a rule forbidding conscious racial discrimination simply gives *carte blanche* to the operation of unconscious, stigmatizing racial stereotypes. But a rule permitting the use of non-invidious racial preferences that is narrowly tailored to promote racial integration will reduce the operation of invidious stereotypes, and is unobjectionable on any coherent account of what is wrong with racial discrimination.[15]

## What are the most important problems in normative ethics and what are the prospects for progress?

Two large-scale problems about the structure of values and normative judgments lie at the core of the most exciting current work in normative ethics. The first problem concerns the sources of norms for warranted emotions and attitudes. Consider a basic emotion such as fear that we share with animals. Untutored fear responds not to the fearsome (that which merits fear), but to what we may call the "scary"—that which has sensory properties that cause fear. Thus, children are often afraid of the dark, which is scary, but not truly fearsome. As children mature, they learn to control their fears somewhat in response to norms for warranted fear. Some scary things, like the dark, they learn not to be afraid of. They also learn to fear things that, lacking sensory qualities, are not scary, but are fearsome—for example, impending insolvency. Value judgments, embodied in norms, help guide our fears to warranted objects. What is the origin of these norms, and how can they be justified? They don't have any simple connection to other considerations, such as morality or self-interest.[16] Although it is plausible to suppose that fear evolved to protect animals against threats to their health, bodily integrity, or survival, norms for warranted fear don't track these dimensions of well-being in any straightforward way. It would not be prudentially rational to fear a vicious dog who "smells" fear. Yet the dog is genuinely fearsome—it *merits* fear. Reasons of self-interest are reasons of the wrong kind to dictate the shape of the fearsome, even if they override considerations of what to fear in particular

---

[15] "Racial Integration as a Compelling Interest," *Constitutional Commentary* 21 (2004): 101–127; "Integration, Affirmative Action, and Strict Scrutiny," *NYU Law Review*, 77 (2002): 1195–1271.

[16] Justin D'Arms and Daniel Jacobson, "Sentiment and Value" *Ethics* (2000) 110: 722–48.

cases. What are reasons of the right kind, and how do we justify their intrinsic connection to fear? This remains one of the outstanding problems of sentimentalist theories of value such as my own.[17] Research on this problem is active, but I can't say now what the ultimate shape of a solution to it would be.

The second large-scale structural problem in normative ethics concerns a relationship of consequentialist value judgments, which apply to states of affairs ("This state of the world is better than that"), to deontic, agent-centered requirements ("Each person should keep her promises"), which apply to agents. If we conceive of the point of action in consequentialist terms, as promoting good states of affairs, it is difficult to justify agent-centered constraints.[18] If breaking promises is bad because doing so disappoints people, it may be the case that by breaking a promise I made to A, to A's disappointment, I can spare B, C, and D greater total disappointment. Yet commonsense morality does not allow the fact that others would be disappointed if I kept my promise to A to excuse me from keeping my promise. Nor am I excused if, by breaking my promise to A, I prevent three other people from breaking their promises to B, C, and D. In the 1980s and 1990s, it seemed that the most promising avenue for justifying agent-centered moral constraints lay with neo-Kantian strategies, which attempt to derive such constraints from Kant's Categorical Imperative. However, serious difficulties arose both in showing how to apply the categorical imperative so that it gave us intuitively plausible moral requirements for the right reasons, and in justifying the categorical imperative along anything like Kantian lines. In recent years, Stephen Darwall has laid out an entirely new way to approach the problem, based on what he calls "the second-person standpoint."[19] This is the perspective we take up whenever we address demands on others to act in certain ways -for instance, to keep their promises to us, or to respect our privacy. In making such demands, we presuppose our own authority to prescribe norms, and hold others accountable for following them. Darwall persuasively argues that the domain of justice, and of moral

[17] For further discussion, see Wlodek Rabinowicz and Toni Ronnow-Rasmussen, "The Strike of the Demon: On Fitting Pro-Attitudes and Value" *Ethics* (2004) 114:391–423.

[18] Samuel Scheffler, "Agent-Centered Restrictions, Rationality, and the Virtues" *Mind* (1985) 94:409–19.

[19] Stephen Darwall, *The Second-Person Standpoint: Morality, Respect, and Accountability* (Cambridge, Mass.: Harvard University Press, 2006).

rightness more generally, along with associated attitudes such as respect, outrage, resentment, and guilt, essentially spring from this second-person point of view. They cannot be grounded in the impersonal third-person observer's point of view that structures consequentialist judgments of the value of states of the world. This is a breakthrough insight, and provides the most promising avenue today for explaining the relationship between consequentialist and deontic moral judgments. Each type of judgment springs from a fundamentally distinct point of view, each tied to a different kind of normative force: while others can hold us accountable for demands validated from a second-person standpoint, they cannot hold us accountable for our failure to make the world as good as possible. Darwall's account offers an exciting new way forward for deontological moral theories.

# 2

# Roger Crisp

## Uehiro Fellow and Tutor in Philosophy
St. Anne's College, Oxford, UK

---

**Why were you initially drawn to normative ethics?**

Between 1968 and 1979 I attended Brentwood School. The school had high academic standards and some outstanding teachers. I'd single out Dennis Riddiford among them, a very fine classical scholar. At his retirement dinner, Dennis said he'd seen his role as Promethean—to bring the fire he found in classical texts into the lives of his students. And that's how it was. My interest in ancient ethics, and hence contemporary ethics, began with his lessons.

The school was rather old-fashioned. In the junior school, one could be beaten for speaking during a meal after a bell had been sounded. Hair had to be kept above the collar. There were many other rules, with varying sanctions available to teachers and 'praeposters' to enforce them. Living under such a regime made me think quite hard about the role and function of norms in our lives. At times, particularly perhaps in my early teens, my position veered close to anarchism.

In my later years at the school, I began to take a greater interest in politics. I witnessed the rise of Mrs. Thatcher – the 'Iron Lady' – and her frequent use of cold war rhetoric resonated with me. I knew quite a lot about the military power of the Soviet Union, and I was afraid of it. I was certainly no libertarian and was quite aware of the many imperfections in western 'liberal democracies', but I did at that time begin to see the value of individual and political freedom. The 1970s were something of a high-point in the UK for racist political parties such as the National Front, and I became involved in anti-racist movements such as the Anti-Nazi League. At that time, I suppose I would have held to some kind of 'equality of respect' position, one I later extended to include non-human animals.

Most of the reading I did around these topics was political or historical. My intellectual horizons were largely shaped by the subjects I studied at school, though I became very interested in the visual arts through the influence of my parents. Philosophy was not, and on the whole still is not, a school subject in the UK. I think that, especially given our philosophical heritage, that is a real pity, and the complaints I have heard about school philosophy strike me as primarily about the way that the subject is taught. The first philosophical work I studied in any great depth was Plato's *Phaedo*, at Oxford. I found it quite hard going, but fascinating and my excellent tutor Gwynneth Matthews was a great help. Then I went on to Locke's *Essay*, which nearly put me off philosophy for good. I couldn't understand what questions Locke was asking, and his answers seemed to me rather 'made up' (I think I had slightly verificationist leanings). It wasn't until reading Mill's *Liberty* that I found a real connection with philosophy. It was immediately obvious to me how many of the things Mill was arguing for chimed with conclusions I'd arrived at myself, in, of course, a much less rigorous way. Indeed it was something like the 'harm principle' that I'd sometimes advocated at school in lively discussions with my headmaster.

By this point I was fully hooked on philosophy, especially moral and political philosophy, and I read as much as I could, from the *Republic* to Rawls and beyond. I remember several heady days in Kew Gardens immediately after my final exams, working through Nagel's *Mortal Questions*. I managed to land a place on the Oxford Masters programme in philosophy, the B.Phil., and had as my main supervisor the inspiring John Ackrill, who taught me the Aristotle paper. I was hugely lucky with my graduate supervisors at Oxford. My other B.Phil. supervisors were Jonathan Glover (for my thesis in philosophy of mind and language), James Griffin (moral philosophy), and Alan Ryan (political philosophy). I went on to write a doctoral thesis on ideal utilitarianism, supervised mainly by Griffin, but also by Michael Lockwood, Derek Parfit, Joseph Raz, and David Wiggins. At the end of the 1980s, jobs in philosophy were pretty scarce, and I did consider other careers, including the law. But I hung on for a year or two with some part-time temporary posts, and was then awarded a British Academy Post-doctoral Fellowship, which I held at University College. My college room had been J.L. Mackie's, and that next door was shared by Wiggins and Ronald Dworkin. I also had the opportunity to get to know H.L.A. Hart well, as well as many other distinguished philosophers involved with University College. I was then offered the post I now hold at St Anne's College, an institution with important links to normative ethics. The College emerged out of the Association for the Education of Women, of which T.H. Green was the first secretary (indeed, I write from what would have been T.H. Green's bedroom, had he not sadly died before being able to move into his new house). Former fellows include Iris Murdoch and Gabriele Taylor.

**What example(s) from your work (or the work of others) illustrates the role that normative ethics ought to play in moral philosophy?**

I'm sure there are lots of roles normative ethics ought to play in philosophy. But first I need to say a little about how I draw the boundaries around normative ethics itself. It's become quite standard to separate off normative, or 'first-order', ethics from metaethics, which is seen as a second-order enquiry *about* ethics rather than within it. So normative ethics might include, say, elucidation of utilitarianism and discussion of abortion, but not issues concerning the truth-aptness of moral judgements or the metaphysical status of moral properties.

One of the many things I've learned from Jim Griffin is to be suspicious of distinctions in philosophy, and this is a case in point. Take someone like Mackie. Mackie argued that there are no moral properties such as rightness or wrongness. An ontology which included them would be just too 'queer' to believe, nor is there any plausible story to be told about how we could detect them. Now, if we apply something like Russell's theory of descriptions to certain moral propositions, they turn out to be false. Take, for example, 'Abortion is wrong'. This states that there is a property – wrongness – which abortion has. Since there isn't such a property, the proposition is false. And the same would go for 'Abortion is right', 'Abortion is permissible', and so on. This didn't worry Mackie too much, as he was happy to isolate metaethics from first-order ethics, and so felt able to continue making moral judgements, such as that the death penalty is always morally forbidden. But there seems something paradoxical in claiming both 'The death penalty is wrong' and 'The proposition "The death penalty is wrong" is false'. One could, of course, try to argue that moral judgements aren't what they seem, and are, say, expressions of some kind. But that kind of reinterpretation has never seemed to me at all plausible.

Nevertheless, there clearly is a spectrum here, to be understood in terms of directness. The primary question in normative ethics, I suggest, is: 'How should I act?'. (I used to think it was 'How should I live?', but Michael Smith persuaded me that everything important for ethics in this latter question could be captured in the former.) If one arrives at an answer to that question which goes via issues in philosophy of language or metaphysics, it will be a pretty indirect route, and the chances are that one won't be doing normative ethics. But if one takes the question head on, and develops an answer relying, say, on an account of the virtuous agent, then that is normative ethics.

One primary role of normative ethics, then, is to clarify the different options available in answering the question about how we should act, and to provide arguments for and against them. It's become quite common now to claim that there are essentially three types of ethical theory: consequentialist (or utilitarian), deontological, and virtue ethical. This is another distinction I'm suspicious of. First, it's very tricky actually to come up with a definition of one of these categories which includes just what one wants and no more. Take deontology. The Greek word *deon* means 'one must', and so we might call a deontologist someone who thinks that there are certain things, such as killing the innocent, which one must not do. But utilitarians – usually seen as paradigm non-deontologists – think there is a certain thing which is absolutely forbidden, *whatever the consequences* (e.g., even if lots of people end up being tortured): not maximizing the overall good. Likewise, a virtue ethicist may claim that one must, in all circumstances, do the action a virtuous person would do, adding perhaps that a virtuous person would never kill the innocent.

Second, this tripartite distinction seems to distort the philosophical tradition. Aristotle is usually seen as the paradigm virtue ethicist. But, like the utilitarians, he puts great weight on the idea of happiness and its promotion. And like many deontologists, he believes there are certain rules which can guide our actions, such as that we should render honour to older people, by giving up our seat to them and so on.

As I've said, I think meta-level inquiries into the nature of ethics can have normative implications, and I've written a little on that. I've argued, for example, in favour of a non-naturalist form of realism, which avoids any argument that naturalism in ethics leads in the direction of nihilism, and also for a form of 'generalism', which allows ethical principles to be rational and explicable.

I've also written a certain amount in the history of moral philosophy, especially on Plato, Aristotle, Hume, Mill, and Sidgwick.
I think such discussions do have independent historical interest,
but the main purpose of the history of moral philosophy, as I see
it, is again to assist us in answering that primary normative question about action. By interpreting some past philosopher, we may
uncover options and arguments which would not otherwise have
occurred to us, and that's certainly how it has seemed to me in
my work in this area. Nor would I want to claim that my interpretations are clearly right and those of others clearly wrong. This is
not just because I'm a Pyrrhonist (more on that later). It seems
to me that philosophical texts, like other texts, are indeterminate,
such that often a large variety of competing and roughly equally
plausible interpretations is available. This is why excessive clarity
or precision can be philosophical vices, if they result in a text's
being less suggestive than it might otherwise have been.

I've also tried to clarify and provide arguments for and against
various positions in normative ethical theory. I sought to clarify
the most plausible form of utilitarianism in a 1992 paper called
'Utilitarianism and the Life of Virtue' and in some chapters of
a book on J.S. Mill. In recent years, I've moved away from the
more 'objective' conception of well-being I advocated in that Mill
book, in the direction of hedonism, the view that pleasure is the
only good (see my paper 'Hedonism Reconsidered' or the slightly
more developed chapter in my book *Reasons and the Good*). My
view on the rise of virtue ethics in the 1980s and 1990s was that
it was resurrecting several important questions about the nature
of ethics, such as the importance of character and motive, or the
demandingness of morality, which had to some extent fallen by
the wayside. So I've tried to say a little about how it's done, for
example in an introduction to an edited volume of specially com-
missioned essays on the virtues. Thinking about demandingness
and the nature of the self led me away from impartialist ethics in
the direction of something more like Sidgwick's 'dualism of practical reason'. I wrote a paper on that in 1996, and have developed
the view further in a chapter of *Reasons and the Good*. I've also
made contributions to political philosophy, concerning the nature
of liberalism and the best theory of equality (I argue for 'sufficien-
tarianism', the idea that justice requires that each individual has
enough for a good life), and also to applied ethics, including the
issues of killing and euthanasia, advertising, the moral status of
non-human animals and the environment, the allocation of health

care resources, informed consent, and genetics.

## How do studies within scientific disciplines contribute to the development of normative ethics?

The question implies that normative ethics isn't itself a science. That's something I find rather dubious, and I'd prefer to see ethics alongside the other disciplines that seek to tell us what the world is like. That's not to say that I'd want to deny some kind of fact/value distinction. But I do think there are facts *about* values (and reasons), and that discovering these is the aim of normative ethics.

Which other sciences or disciplines, then, can helpfully contribute to normative ethics? One obvious one – the significance of which Bernard Williams stressed especially in his final years – is history. Normative ethics, because it uses language, concerns the manipulation of concepts, and each of those concepts – 'ought', 'wrong', 'forbidden', 'rights', 'justice', 'mercy', 'cruelty' – has its own history. Of course, we want to avoid the fallacy of thinking that the meaning of a term can be found in its etymology. But we can learn a great deal about a concept by understanding the historical and cultural context in which it emerged, and how it has changed. Consider, for example, the concept of '*to kalon*' or 'the noble' in Aristotle's *Ethics*. Here the disciplines of ancient history and classics provide invaluable insight into that notion, and, as I've argued in some papers on Iris Murdoch and on Aristotle, it is a notion which remains part of our contemporary lived morality but is largely ignored by philosophers. Sociology can also provide insight into moral concepts. Consider here Peter Berger's path-breaking work on the concept of honour, or the pervading influence of sociology on the thinking of Alasdair MacIntyre.

Morality itself is a social phenomenon with its own history, and a discipline from which philosophers could learn a good deal more is social anthropology. I myself have found quite persuasive the work of authors such as Christopher Boehm, who claims that morality developed into its current form about 100,000 years ago. His argument is based partly on primatological data, but also on evidence concerning existing bands of hunter-gatherers. I have come to believe that morality has developed essentially as a form of social or group control, the function of which includes the avoidance and resolution of conflict, and the solving of co-operation problems. I take morality to include not only the various action-guiding norms and ideals that we live by, but also the emotional 'sanctions' of shame, guilt, and blame. Once we see morality for what it is, rather than ourselves manipulating its concepts, as most philosophers continue to do, we can step back from it and try to understand whether we have any reason to continue with it. I have tried to begin that task in the first chapter of *Reasons and the Good*, where I suggest also that there are strong analogues between morality and legal systems. So another route into understanding morality and its concepts will be via jurisprudence.

Another source of exciting insight into the nature of morality in recent years has been psychology, broadly understood. Work using fMRI by the cognitive neuroscientist Joshua Greene at Princeton, for example, has suggested that some types of moral judgements involve more 'rational' elements of the brain, others more 'emotional'. The evidence itself is, of course, itself contestable, and the interpretation of it can go either way. Only in the last day or two, the UK newspapers have included articles about work at Harvard and elsewhere on the ventromedial prefrontal cortex, which is said to show that utilitarian judgments are more likely to be made by those with damage to their brains, because they lack the normal emotional 'block' to such decisions. The articles I have seen all conclude that the utilitarian decision is thereby thrown into doubt, but this is to assume that the emotions help us to understand the world. But why should we accept this? The main role of the emotions in evolution has been to help us survive, not to understand normative truth. But it is not only neuroscience that has been providing material for normative ethics. Recent work by Stephen Stich, John Doris, and others has brought out very clearly how information from empirical psychology more generally may have implications for our understanding of the nature of moral motivation, the virtues, and much else. And the work of Ed Diener, Martin Seligman, and others in 'positive psychology' has huge practical significance, at least some of which should be reflected in philosophical enquiries into well-being and in applied ethics more broadly.

Finally, what of art and literature? Iris Murdoch claimed, and she is followed these days by Martha Nussbaum and others, that the appreciation of art and literature provides a surer route to moral insight than mere philosophy. I think that's unlikely. It all depends on the kind of philosophy you are attracted to. In the case of Murdoch and Nussbaum, it's clear that their reflections upon art and literature have been highly fruitful. But that's not to say that there is anything wrong with the generally non-literary style of most contemporary normative ethics. If one's developing a sophisticated form of utilitarianism, constant references to Sophocles or Dickens will probably be more of a distraction than anything else. And there seems to me no evidence to suggest that those well versed in literature are more likely to capture the truth about ethics. What's important is that people go about the job as well as possible, with the tools best suited to their own purposes. There are, as we say in the UK, many ways to skin a cat.

### What do you consider the most neglected topics and/or contributions in normative ethics?

I've already mentioned the idea of the noble or the fine. It is absolutely central to Aristotle's developed normative ethics, which makes it all the more surprising that there has not been more discussion of it in recent philosophy. For most virtue ethicists are happy to be identified as neo-Aristotelians. According to Aristotle, the virtuous person – the moral agent, that is – acts for the sake of the noble. The noble is the highest good, to the point that the virtuous person would rather die than do something ignoble, such as run away in battle: 'the good person is seen to assign himself the larger share of what is noble'. Nobility is closely tied to Aristotle's key virtue, magnanimity or greatness of soul, which itself has received surprisingly little attention in recent literature. The great-souled person seeks honour, and in the dependence of that good on the opinions of (even if only idealized) others, we see an aestheticizing move in Aristotle which is contrary to contemporary sensibilities. But there is undoubtedly still a large place for what one might call 'moral value' in our everyday moral framework. There is something admirable in certain actions of certain people – such as those who risk their lives to save others, or tirelessly work to decrease suffering – and this value is closely related to Aristotelian nobility. A major question in ethics is whether there really is such value, or whether it is merely an epiphenomenon of the 'morality system' that has emerged in the way I described above as a means of social coercion. If we decide that there is, then the prospects for so-called 'welfarist' theories such as utilitarianism, according to which the only value is well-being, will appear much bleaker.

There is plenty of writing in virtue ethics at the formal level, which seeks to show in broad outline what is wrong with utilitarian or Kantian views and what a proper virtue-based theory would look like. But oddly there is little discussion of which human traits are virtues and vices. That again contrasts starkly with Aristotle. Most modern writers on virtue ethics, when seeking guidance from Aristotle, jump from the middle of book three of the *Nicomachean Ethics* to the beginning of book six, missing out Aristotle's detailed and insightful discussions of courage, generosity, justice, and much else. The same goes for the important writings of Hume on particular virtues and vices. I suspect that there at least two reasons for this. First, philosophers now work primarily within the academy, writing for other philosophers, and most of them have little intention actually to change the lives of other people, though I suppose they might quite like it if they did. Philosophers, that is to say, are more interested in broad philosophical theories than practical detail. Second, many of the writers who offer detailed accounts of particular virtues – Plato, Aristotle, Aquinas, Hume – are hoping, through their descriptions of the lives of those with these virtues and their corresponding vices, to persuade us that the virtuous life is the life that is best for the virtuous person. So it can never be rational to choose the vicious action. That's a view very few people are now prepared seriously to entertain, which is a pity, since this is a major part of what gives excitement and life to ancient ethics.

Another position widely discussed in the ancient world and now largely neglected I have already mentioned: hedonism. These days hedonism is usually quickly dismissed in philosophical writing and lecturing, with a brief reference to Robert Nozick's famous example of the 'experience machine'. According to hedonists, my life could go as well for me if I were hooked up to an experience machine, giving me experiences with the same content, or apparently the same content, as those I am having in the 'real world'. So hedonism is false. This is another case where some conceptual archaeology is called for. What is the basis for our rejection of the experience machine? Why are we so attached to 'authenticity' and does it really matter? Whatever one's view of the plausibility of hedonism, these are questions that a normative ethicist should not ignore.

**What are the most important problems in normative ethics and what are the prospects for progress?**

As I've said, the most fundamental question for normative ethics is how we should act. That 'should' isn't 'sub-scripted' in any way.

The question is not, say, how *morally* we should act. Theories in normative ethics are best understood as accounts of *reasons* to act in one way or another, so to this extent the concept of 'reason' is itself a foundational one. In normative ethics over the last thirty or so years, there has been a fruitful turn in the direction of reasons, in the work of, for example, John Broome, Jonathan Dancy, Susan Hurley, Thomas Nagel, Derek Parfit, Joseph Raz, Tim Scanlon, John Skorupski, Bernard Williams, and others. Good progress has already been made, and some consensus is developing around the idea, important early in the work of Raz, that reasons are to be understood in terms of how they favour or disfavour actions. Exactly how the favouring relation works, however, as well as what reasons we have and how they are to be weighed against one another, are issues on which there is as much disagreement as there has ever been in ethics.

So here the general issue of methodology arises, and in particular that of the epistemology of normative ethics. If there are truths to be had here, how are we to discover them, and how are we to understand ethical knowledge? Should it be on some kind of foundationalist model, as in the recent work of intuitionists such as Robert Audi, much of which I find myself in broad agreement with? Or should we advance along coherentist lines, as, for example, in the work of David Brink? Here there remains much to be done in exegesis of the alternatives and development of the arguments for and against them.

One aspect of normative ethics in particular wins the title for largest elephant in the room: the breadth and depth of disagreement within the area. There are broad schools – the utilitarians, the Kantians, the contractualists, the virtue ethicists, the rights theorists, the feminists, and so on – but even within these schools there are major disputes about fundamental issues. And of course the schools themselves are in a constant war of all against all. That can make philosophical life somewhat nasty and brutish, of course, but more importantly it raises the question how a philosopher should respond. In recent years, I've come to accept the Pyrrhonist position that what's called for is suspension of judgement, unless one can come up with a plausible story about why those who disagree with one ought not, in the case in question, to be trusted. Since we know so little about the formation of ethical beliefs and the origin of disagreement, it seems to me that those in normative ethics who sincerely believe that they are onto the truth and their opponents quite off the track are guilty of a high degree of epistemic irresponsibility. I have some hope that the non-ethical sciences I mentioned above, and in particular psychology, may over the next few years offer further explanation of the origin of our beliefs, so that in the light of that new knowledge we can amend our views appropriately. But it isn't just up to the psychologists to speak. The philosophers must listen!

# 3

# Gerald Dworkin

Professor

University of California, Davies, USA

---

**Why were you initially drawn to normative ethics?**

I became a moral philosopher because my mother was a communist. I grew up in Manhattan as a "red-diaper" baby. This is a term for people whose parents were members of the Communist Party. In my case it was one parent. My father was a-political but my mother was a dedicated member of the party and remained so until her death.

One consequence of this was that at an early age I began reading Marxist classics on dialectical and historical materialism. As a high school student I attended classes in these subjects at the Jefferson School of Social Sciences which was a Marxist adult education institute in New York City associated with the Communist Party, USA. Among the faculty were a number of leftist academics dismissed from the City University of New York, including the school's director, Howard Selsam. It had as many as 5000 students enrolled per term, but the Subversive Activities Control Board forced its closing in 1956.

Selsam was a serious philosopher who had his Ph.D from the Columbia University.

Although I abandoned belief in the philosophical system, as well as its political consequences, soon after I entered the City College of New York in 1955 the interest in philosophical ideas, and in particular normative notions, was the lasting legacy of my childhood upbringing. In fact, my first published article was a critique of dialectical materialism in a journal called Studies on the Left (whose contributors included young scholars such as Eugene Genovese and C. Van Woodward).

Although I was greatly interested in philosophy I decided not to major in it as it seemed very unlikely that one could make a living as a philosopher. Instead – after one disastrous semester as an engineering student – I became a mathematics major. I did, however, take an honors minor in philosophy writing a thesis on the role of theory in science.

Still hedging my bets I decided to enroll in the fledgling logic and the methodology of science program at Berkeley rather than the philosophy program. But after two years I realized my heart was in ethics not mathematics and transferred into the philosophy department. I did get a masters degree in mathematics as a consolation prize with an oral exam by a committee including Addison and Chang. Its most amusing feature was being asked what the fundamental theorem of the calculus was—to which my answer was "I have no idea." Since this answer was true, I passed!

I was fortunate at Berkeley to be there for the three years that Tom Nagel was an Assistant Professor. He supervised my thesis – mostly long distance – since I spent one year as a Woodrow Wilson Fellow in London and then finished the thesis while I was an instructor at Harvard. The thesis was on the nature and justification of coercion and it contained the first draft of my article on paternalism— the most frequently reprinted article of mine. The general topic of coercion, particularly the question of what conduct may be legitimately criminalized, has remained a constant theme in my work.

## What example(s) from your work (or the work of others) illustrates the role that normative ethics ought to play in moral philosophy?

If one thinks of moral philosophy as being divided between meta-ethics and normative ethics then there is no particular role that normative ethics plays; it simply is (one) part of moral philosophy. It is interesting historically that when I was in graduate school there was great hostility to thinking of normative ethics as part of philosophy at all. It was thought of as "preaching" rather than philosophy. Only meta-ethics was considered "real" philosophy. To those who have entered philosophy since the publication of John Rawl's *A Theory of Justice*, and the founding of the journal *Philosophy and Public Affairs*, the suspicion about normative ethics will seem quite incredible. Of course, even at the time the critics had to dismiss large parts of Plato, Kant and Mill as simply their views as persons or citizens, not their views qua philosophers.

It may, in fact, be time to re-introduce a certain amount of humility about the authority of moral philosophy to make normative judgments. One might regard Mill' statement that people '... must place the degree of reliance warranted by reason, in the authority of those who have made moral and social philosophy their peculiar study... [R]eason itself will teach most men that they must, in the last resort fall back upon the authority of still more cultivated minds' as a piece of Victorian authoritarianism. My own view is that there is something both gained and lost as a result of philosophical training with respect to being reliable on moral issues. What is gained is that we have much experience in reading, thinking and writing about these issues. We are trained to evaluate and criticize arguments. Our views have been exposed to critical examination and refutation by other philosophers. The downside is that – either as a result of the training or because those who go into philosophy tend to have certain personality characteristics – we have a tendency to accord more weight to argument than to sympathetic feelings, experience with the subject matter, intuitive insight, and so forth. If we re-formulate the question as what role normative ethics ought to play in meta-ethics then I think there are a number of points to be made.

First, it can provide data to test various meta-ethical views. Our normative views on various issues provide checks on meta-ethical views not in the sense that they are the last word but as the first word. Meta-ethics exists to provide explanation and validation of normative ethics. Second, normative ethics can give us a rich sense of the variety of ethical questions and modes of reasoning that are part of our practical reasoning. If, for example, deductive reasoning from exceptionless principles is not the way that most of us reason about normative issues then meta-ethical frameworks must either adjust to accommodate that fact, or proved some kind of error theory as to why most of us reason the way we do. It is also possible, as Bernard Williams thought, that our best understanding of normative ethics will not come from any form of highly abstract, philosophical theorizing but from a better understanding of the historical context in which we think about normative problems, e.g. what counts as a normative problem and what constitutes a normative solution to that problem.

**How do studies within scientific disciplines contribute to the development of normative ethics?**

I think that the most interesting connection between normative ethics and other disciplines is from the latter to the former. If one accepts some version of the "ought implies can" principle then various disciplines may provide bother particular facts and theoretically derived constraints on what kinds of normative demands are possible for creatures such as ourselves. For example, evolutionary biology cannot tell us what practices or institutions are justified or right, but they might be able to give us suggestions about what kinds of normative systems are easier or harder for us to adopt and live by. This would not tell us that such a normative system is "incorrect" but it would give us information about the costs of it being the correct one.

Again, history or sociology might be able to give us information about what kind of social system precedes (and conceivably make possible) changes in moral standards and norms. Why are doctors (on the whole) less paternalistic towards their patients in some cultures rather than others? One contribution going in the other direction has been the influence of normative theories of distributive justice on the choice or problems and the evaluation of solutions in economics. The tool of the "veil of ignorance" has been widely used by economists in the design of experiments in behavioral economics. The use of other standards than pareto optimality to judge social welfare functions has been due to normative theory. Similarly for criticisms of cost-benefit analysis in public policy evaluations, particularly in the assessment of environmental changes. Finally, much of legal theory is influenced by contemporary normative theory—not always for the better.

## What do you consider the most neglected topics and/or contributions in normative ethics?

One neglected topic is the possibility of environmental constraints on possible normative systems. Just as evolutionary biology can tell us something about the genetic limitations of ethical systems there ought to be something that the social sciences can tell us about the environmental limitations. The common premise is that ethics must be relevant to the type of creature we are, and it is clear that the type of creature we are is partially determined by the types of environments we face. So, there are design requirements for possible moral systems and it would be interesting to investigate what these are.

Another topic that has been neglected is the pragmatics of moral statements. The pragmatics of a body of language concern how people use a language to accomplish certain effects, or what people must presuppose, or assume, in order to make sense of what other people say. Almost all recent moral philosophy has focused on the syntax or the semantics of moral discourse. Discussions of moral realism, or prescriptivism, or projectivism, or naturalism, have concentrated on the syntax (are moral statements disguised imperatives?) or semantics (do they have truth value?).

I also belive but with a few exceptions – Gerald Cohen and myself come to mind – there has been no attention paid to the ways in which a moral judgment can be inappropriate although true. An example would be one burglar saying to his confederate 'You are doing something immoral." This statement is certainly true but something about the situation makes it infelicitous (as Austin put it) for the burglar to say it. Not everyone has the moral authority, or is in a moral position, to make a (correct) moral judgment.

My methodological hypothesis is that looking at the pragmatic aspects of moral discourse may throw light on various moral phenomena that remain unilluminated by the exclusive concentration on syntax and semantics. I also believe that some of the recent work in the area of neurobiology and ethics is work that ought to be neglected. It is too crude, reductive and simplistic to be worth doing. One cannot dismiss Kant because some part of the brain (associated with emotional response) lights up when people think about trolley problems.

## What are the most important problems in normative ethics and what are the prospects for progress?

It seems to me quite impossible to determine in advance what the most important normative problems are and what are the prospects for progress on these problems. I can tell you what I find interesting and important but only history will pass its judgment on whether those problems turn out to be important. Philosophical problems in general tend to ripen, multiply, transmute, get taken over by other disciplines, disappear, re-appear, all without obvious reasons and causes. Who could predict looking at Plato that the social contract would be still going strong as a philosophical theory, or that the immense attention paid to fine distinctions in ordinary language in the 1960's would disappear from the research agenda? I think moral philosophy is still at the stage where we can only hope that we know less as time goes on.

# 4

# Fred Feldman

## Professor of Philosophy

University of Massachusetts at Amherst, USA

---

**Why were you initially drawn to normative ethics?**

I was drawn into normative ethics more or less by accident. Perhaps I should say "by accidents", since there were two instances in which I became interested in questions in normative ethics in connection with some work I was doing about topics that are not in normative ethics. In each case I thought I would say a few things about some dispute in normative ethics merely to clear up some questions that were left over from some previous work. The "clearing up" operations have been going on for about twenty-five years.

The first accident occurred in connection with some ideas I had in deontic logic. As a graduate student, I had the honor and pleasure of being able to work under the direction of Roderick Chisholm. At the time I was most interested in some questions in metaphysics and in the history of philosophy. My dissertation concerned metaphysical and logical questions about various non-strict conceptions of identity. I had no reason to suppose that I would turn my attention to ethics.

However, Chisholm had posed an interesting and pretty tricky question about the logic of obligation.[1] The puzzle concerned the logical relations among four sentences:

1. Jones ought to go to the aid of his neighbors.

2. If Jones goes to the aid of his neighbors, he ought to tell them he is coming.

---

[1] In his 'Contrary to Duty Imperatives and Deontic Logic', *Analysis* 24 (1963): 33-36.

3. If Jones does not go to the aid of his neighbors, then he ought not to tell them he is coming.

4. Jones does not go to the aid of his neighbors.

These seem to be consistent – such a collection of sentences could be true of some actual person. Furthermore, none of them seems to entail any other. But what, precisely, is their logical form?

Using '$O$' as an obligation operator, and using '$G$' and '$N$' to indicate *Jones goes to the aid of his neighbors* and *Jones notifies them that he is coming* and using '$\rightarrow$' to indicate some sort of conditional connective, we might try to represent these four sentences as:

1a. $OG$

2a. $G \rightarrow ON$

3a. $O(\sim G \rightarrow \sim N)$

4a. $\sim G$

But, as Chisholm pointed out, this will never do. For 1a and 2a seem to entail $ON$, and 3a and 4a seem to entail $O \sim N$. In other words, when so formulated, the sentences are not consistent. They entail a serious conflict of obligation. They entail both that Jones ought to tell his neighbors that he is coming and that Jones ought not to tell them that he is coming.[2] Chisholm went to discuss other possible ways of displaying the logical form of sentences (1)–(4). None of these satisfactorily captured the ideas intended by the original sentences, since none of them preserved the consistency and independence of the originals.

I was fascinated by this puzzle. I worked on it for several years on and off. Eventually I wrote some papers in which I defended an informal system of deontic logic that had the capacity to express those sentences in a form that would preserve and clarify their logical structure.[3]

The semantics for this theory involved certain assumptions: that there are possible worlds; that these worlds can be ranked for some relevant sort of value; that worlds are accessible to agents

---

[2] If you assume that O∼N entails ∼ON, then you can derive an outright contradiction from the sentences under this formalization.

[3] The system of deontic logic was eventually published in my *Doing the Best We Can: An Essay in Informal Deontic Logic*, Reidel, 1986.

at times; that obligation statements must be relativized to times. I maintained that our fundamental obligation as of a time is to behave as we do in the best of the worlds then accessible to us. Using '$s$' to indicate a person, '$t$' a time, and '$p$' a state of affairs, I expressed by saying '$MOs, t, p$'. I also maintained that a statement of conditional obligation ('$MOs, t, p/q$') should be understood to mean that $p$ occurs in all the best $q$-worlds accessible to $s$ as of $t$. I proposed that the Chisholm sentences should be understood in this way:

1f. $MOj, t, g$

2f. $MOj, t, n/g$

3f. $MOj, t, \sim n/ \sim g$

4f. $\sim g$

I visited various departments, presenting talks on this proposed way of formalizing the logic of obligation. In the question periods following the talks, I was often asked about the ranking of possible worlds for "some relevant sort of value". How, precisely, did I think the worlds should be ranked? Did I want to say that the value of a world is just the sum of the values enjoyed or suffered by the inhabitants? And if so, what did I want to say about worlds in which the value was distributed in a very unfair way? So, in order to be prepared for questions like these, I began to think about the ranking of possible worlds.

Thus, largely as an afterthought arising in connection with my work in deontic logic (which I take to be in metaethics), I was drawn into a consideration of one of the central questions in normative ethics – what makes one possible world better (in the relevant way) than another?

Another line of questioning often arose in question periods. Colleagues would point out that the conception of obligation expressed by my '$MOs, t, p$' is in many ways similar to the utilitarian conception of obligation according to which one morally ought to perform an action if and only if it maximizes utility. They would then ask (often with the now-famous incredulous stare) 'do you really mean to defend act utilitarianism? Don't you realize that it has been utterly abandoned? What are you going to say about justice and integrity and personal projects? What are you going to say about duties of special obligation?'

It was obvious that I would have to come to these talks prepared to answer questions like these. I hoped at the time (and continue to hope) that it will be possible to adjust the ranking of accessible worlds in such a way as to take appropriate account of justice and integrity and duties of special obligation. As I tried to work out the details of the neo-utilitarian theory, I became more deeply drawn into another question in normative ethics: what makes for the moral rightness of actions? I found myself adjusting and revising my conception of value so as to make my theory of obligation yield more palatable results. In any case, as a result of questions concerning my views in deontic logic, I found myself pursuing a second central question in normative ethics: what is the criterion of moral rightness for actions?

The second instance in which I was drawn into normative ethics touched me much more deeply and in a more personal way. After a long illness, my daughter Lindsay died when she was about sixteen years old. As any parent would be, I was shocked and dismayed. For a long time I found it difficult to think about anything other than her death. But gradually my thoughts began to turn in the direction of some philosophical questions about death. I read some of the relevant literature. I began to see that many philosophers think that death cannot harm the one who dies. The old Epicurean argument is well known: death cannot harm you while you still live, for it has not yet come. Death cannot harm you after you have died, since at such times you no longer exist and cannot suffer any harms. Therefore, death cannot harm you at any time. This line of thinking seemed to me to be preposterous. I took it to be perfectly obvious that my daughter's premature death was a grievous harm to her.

I began to think more about the harm of early death. I came to the conclusion that death is bad for the one who dies to the extent that the death deprives that person of the good she would have enjoyed if she had not then died. I imagined that if my daughter had not died when she did at age sixteen, she might have had a long and happy life. Her death was a great harm to her, I thought, precisely because it deprived her of this good life. I tried to imagine the value for her of the life she actually lived, with death occurring at age sixteen. I tried to imagine the value for her of the life she would have lived if she had not died at age sixteen. (I assumed that the relevant life would be a life in which she doesn't have a brain tumor, and doesn't undergo extensive surgery and radiation; I assumed that it would be a life in which she enjoys pretty good health for seventy-five or eighty years.) It seemed reasonable to me to say that the evil of her death for her is equal to the difference between the value for her of her actual life and the value for her of the life she would have had if she had not then died.

I wrote some papers in which I developed and defended this view about the evil of death.[4] Again I was invited to give talks at other departments. Lots of interesting questions were raised, but one in particular seemed to keep coming up: precisely how do I determine the value for a person of a life? This is apparently equivalent to the question about individual welfare, or "well-being". So I was facing the question: what makes a person's life go well for her? Or, in somewhat more old-fashioned jargon: what is the good life?

This is a third instance in which I was drawn into a fundamental question in normative ethics as a result of a prior interest in a question not in normative ethics. I was interested in a fundamentally metaphysical question: how can death be a harm to a person if the person does not exist during the time that she is dead?[5] The problem arises no matter how we calculate individual welfare, but my friends and colleagues wanted me to explain my view about how I would calculate individual welfare. This, I take it, is one of the most fundamental question in normative ethics.

Thus, as a result of an interest in a logical puzzle about the concept of obligation (a puzzle I locate in metaethics) I was drawn into two of the central questions of normative ethics: 'what makes right acts right?' and 'what makes one possible world better than another?' As a result of an interest in a fundamentally metaphysical question about the evil of death, I was drawn into another of the central questions of normative ethics: 'what makes a person's life go well for that person?' These are the questions that have occupied me for the past fifteen or twenty years.[6]

## How do studies within scientific disciplines contribute to the development of normative ethics?

We live in an age of science worship.[7] Philosophers often seem to be in awe of scientists. Some, perhaps influenced by Quinean

---

[4] "Some Puzzles about the Evil of Death", *The Philosophical Review*, C, 2 (1991): 205-227.

[5] My views on this question were presented in *Confrontations with the Reaper: A Philosophical Study of the Nature and Value of Death*, Oxford University Press, New York, 1992, xiv + 220 pp.

[6] My views on individual welfare were presented in *Pleasure and the Good Life: On the Nature, Varieties, and Plausibility of Hedonism*, Oxford University Press, Oxford, 2004, xi + 221 pp.

[7] Michael Huemer makes this claim in his recent book *Ethical Intuitionism*,

naturalism, make remarks that suggest that they think that the time has come for us to abandon philosophy. They say things that suggest that they think that traditional philosophical questions fall into two categories: (a) those that are really empirical questions that would be better left to the scientists, and (b) those that are meaningless pseudo-questions that would be better left for the trash bin. I find it astonishing that there are philosophers who seem to espouse this sort of view, but who continue to teach in philosophy departments, and to publish in philosophy journals and to draw their paychecks as professors of philosophy. I can only speculate on how these colleagues manage to carry on. Don't they worry about seeming to be hypocrites? Doesn't it seem to them that their very practice as philosophers clearly demonstrates that their professed views are either obviously false or disingenuous? Perhaps they compartmentalize.

Others hold a more moderate position. They acknowledge that there is an important difference between questions that properly belong in philosophy and questions that properly belong in empirical science. They think that there are legitimate questions worthy of study in both realms. However, these philosophers hold that the results of research in empirical science can have direct relevance to philosophy. They suggest that philosophers ought to be informed about science and ought to take account in their philosophical writings of the results of scientific research. This sort of position is sometimes held in connection with normative ethics.

A good example of this approach is illustrated by a remark made by Daniel Haybron. Haybron is interested in a fundamental question in normative ethics: what is the criterion of human welfare? What makes a person's life go well for that person? He is drawn toward an answer in the eudaimonist tradition. He thinks, that is, that welfare might be tied to the psychological state of happiness. Haybron is aware of the fact that lots of psychologists are currently pursuing empirical research on happiness. Haybron says:

> Philosophical reflection on the good life in coming decades
> will likely owe a tremendous debt to the burgeoning
> science of subjective well-being and the pioneers, like
> Ed Diener, who brought it to fruition. While the psy-
> chological dimensions of human welfare now occupy

---

section 9.4.4, where he cites Peter van Inwagen's views in *An Essay on Free Will* Oxford, the Clarendon Press, 1983.

a prominent position in the social sciences, they have gotten surprisingly little attention in the recent philosophical literature. The situation appears to be changing, however, as philosophers begin to follow the lead of their peers in psychology and other disciplines and examine seriously the psychology of human flourishing.[8]

The claim that interests me is the claim that philosophers working in normative ethics will owe a tremendous debt to empirical researchers in hedonic psychology. In order to test this claim, one would have to review and somehow organize the empirical work being done in the science of subjective well-being (both at present and in "coming decades"). One would then also have to list all the genuinely philosophical questions about the good life. One would then have to review all of this information in order to identify precisely the instances in which results in empirical psychology would have direct bearing on philosophical questions about the good life. I have to acknowledge that I have not done everything that needs to be done in order to test the claim in this rigorous way.

However, I have done a modest and informal review of anthologies, professional papers, and popular discussions of the new science of subjective well-being. I think I am aware of some of the main areas of research. Here is a list of topics that have been discussed in the empirical literature in question:

**A.** Some of the work is in neurophysiology. It appears that some researchers are trying to determine whether there is some brain state, or other physical state, $P$, such that a human being is in $P$ at a time if and only if he is happy at that time. Some of this research has strongly suggested that certain areas of the brain become more active when the subject is happy, satisfied, content, thinking "happy thoughts", or otherwise "feeling good".[9]

**B.** Some psychologists and others have focused attention on a cluster of phenomena concerning the extent to which changes in economic, social, familial, and other factors affect levels of happiness. Some have suggested that each person has a sort of "hedonic

[8] Dan Haybron, 'Philosophy and the Science of Subjective Well-Being' p. 1, forthcoming in a handbook about the science of subjective well-being.

[9] See, for example, Richard Layard, 'Happiness: Has Social Science a Clue?' The Lionel Robbins Memorial Lectures 2002/3 delivered at the London School of Economics, March 2003.

set point" – a level of happiness to which the person will be inclined to return even in the face of presumably important changes in income, or living conditions. An associated thesis is that we are on a sort of "hedonic treadmill". An individual is earning a certain amount of money and is somewhat unhappy; he thinks he will be happier if he could get a raise; he gets the raise; for a short time he is a bit happier; but then his level of happiness returns to his set point; he thinks he will be happier if he can get another raise. The cycle continues.[10]

**C.** Others have done research on the extent to which our assessments of our own life satisfaction can be affected by things that seem not really to be relevant to the quality of our life. Suppose a subject is asked whether he is satisfied with his life as a whole. If the question is asked on a sunny day, or just after the subject has been allowed to find a seemingly lost dollar bill, he will report a higher life satisfaction than he would on a cloudy day, or at a time when he has not found any "lost" dollar bills. Other factors seemingly unconnected with a person's actual satisfaction with life as a whole turn out to affect the responses subjects give. This suggests that our whole life satisfaction ratings are either often mistaken, or else easily affected by things we might have taken to be trivial and irrelevant.

**D.** Others have noted what they call the "peak-end" hypothesis: when looking back at episodes of happiness or unhappiness, our estimates of their magnitude are consistently distorted. We tend to rate them as bigger if they had bigger peaks or ends. There are other fairly consistent distortions in our estimates of happiness levels: when you look forward to a good or a bad thing, you tend to exaggerate its impact on your happiness.

**E.** Some evolutionary biologists have given thought to the question why we are made so as to experience happiness. How could the propensity to happiness – together with the various distortions and systematic errors in its assessment – help to make a person more likely to leave offspring?

**F.** There has also been quite a lot of sociological research into different happiness levels among genders, races, national groups, income groups, age groups, educational groups, etc. So, for example, someone might try to determine whether increasing income

---

[10] For a good introductory collection of papers see Kahneman, Daniel, Ed Diener, and Norbert Schwarz (editors) *Well-Being: The Foundations of Hedonic Psychology* (New York: The Russell Sage Foundation, 1999) 593 + xii.

matches increasing happiness. Some evidence suggests that the answer to this last question is "no". Although income levels in the United States (adjusted for inflation) have grown dramatically in the past fifty years, self-reported levels of whole life satisfaction are no higher now than they were in 1957.[11]

G. There is also quite a lot of medical research on happiness. Are happier people healthier? Are healthier people happier? And what about drugs?

Those are just a few of the areas of empirical research concerning happiness. When Haybron says that moral philosophers 'will likely owe a tremendous debt to the burgeoning science of subjective well-being' I assume he means to be talking about philosophers who are interested in the normative questions about well-being; and I assume that he means more specifically to be talking about people like himself and like me who think that well-being might be determined by happiness levels. Furthermore, I assume that he means to be saying that we who are tempted by eudaimonism would do well to pay attention to the results generated by this empirical research. It very well might prove important to our reflections on the good life.

I have a couple of comments to make about this view:

First, it is not clear to me that any of the empirical research is actually about *happiness*. In many cases, a psychologist will set out to measure something. He might call it 'happiness'; he might call it 'subjective well-being'; he might call it 'whole life satisfaction'. He needs to construct some sort of objective measurement tool. In many cases this will be some sort of questionnaire containing questions such as 'if I could live my life over, I would change almost nothing.'[12] Answers on a battery of questions such as this then yield some numerical "happiness score" or "whole life satisfaction rating". These scores are then compared to some other value – perhaps income, or neurotransmitter level, or age – and then the result is calculated. The research is said to support some claim about the relationship between happiness and income (or whatever).

But it should be obvious that the crucial philosophical question

---

[11] A lot of this research is sketched in Daniel Nettle's *Happiness: the Science Behind Your Smile*.

[12] I didn't make this up. This question was actually included in a happiness measuring tool developed by Ed Diener and some colleagues. See Ed Diener, Robert A. Emmons, Randy J. Larsen, and Sharon Griffin, 'The Satisfaction With Life Scale', *Journal of Personality Assessment, 1985, (49,1)*: 71-5.

was completely overlooked at the outset. In his eagerness to get on with the research, the researcher has simply assumed that answers to the questions on his questionnaire bear on *happiness*. Yet this is utterly unjustified. Surely a person could be very happy, and yet not want to live over the same life he has already lived. And a person could be very unhappy, and yet be content to give it another try. My point should be obvious: *before anyone can start measuring degrees of happiness empirically, he has to be sure that his measurement tool is measuring happiness. Otherwise, he may be measuring the wrong thing – or perhaps nothing of any interest. In this case his results will in fact have no relevance to the question he thinks he is answering.*

My second comment concerns the relation of all this to normative ethics. It sometimes appears that empirical researchers are keen to get started in measuring happiness levels because they take it to be obvious that a person's well-being is determined by his level of happiness. Thus, it has been suggested that governments should employ psychologists to construct tools to measure whole life satisfaction because (a) to be happy is to be satisfied with your life as a whole, (b) your well-being is determined by your level of happiness, and (c) a benevolent government wants to arrange its policies so as to enhance the well-being of citizens.[13]

When I read remarks such as these, I am dismayed and incredulous. Even if empirical research could determine levels of happiness, nothing would follow about well-being. Empirical research cannot establish that happiness makes for well-being. Eudaimonism is not an empirical claim. If some other theory of well-being is true, then measurements of happiness might be completely irrelevant to well-being. In that case, a benevolent government ought to forget about the happiness of its citizens and focus instead on their level of preference satisfaction, or on their level of moral perfection (or whatever in fact does track well-being).

So my second comment is this: *Before anyone concludes that governments ought to adopt policies that will enhance happiness, he has to be sure that happiness bears on well-being. But the question whether happiness is relevant to well-being is a philosophical question, not an empirical one. It cannot be answered by empirical research.*

Although I cannot establish this point here, I am confident that

---

[13] Kahneman makes this suggestion in the Preface to *Well-Being: Foundations of Hedonic Psychology.*

if a thoughtful reader were to reflect dispassionately on the list of empirical topics cited above, he or she will see that none of this empirical research could have any direct bearing on the normative question about the good life. Even if all of the cited empirical research were brought to marvelous fruition, the philosophical question would remain as before: is eudaimonism true? Is the happy life the good life?

My own view is that if there is going to be a debt of gratitude involving philosophers and empirical researchers, it will be owed by the empirical researchers to the philosophers. For the philosophers may be able to help the scientists distinguish among happiness, pleasure, preference satisfaction, contentment, the perfection of our natures as humans, receipt of items from the objective list, whole life satisfaction, and other things that might be the good. And only someone with real insight in normative ethics will be able to determine which of these (if any) is the one that actually makes for welfare.

Finally, I turn to question 4.

## What do you consider the most neglected topics and/or contributions in normative ethics?

I want to focus in this section on a puzzle in normative ethics that seems to me to be both profoundly important and almost universally overlooked or carelessly dismissed. The puzzle can be introduced by appeal to a simple question:

WTD: What should a person do when he wants to do the right thing, but doesn't know what he should do?

I intend that the 'should's in WTD to be understood to be *moral* 'should's. That is, I take the question to be a request for moral guidance, not for prudential guidance. I do not take WTD to be equivalent to the question: 'What would be a prudentially wise thing for a person to do when he does not know what morality requires of him?' I take the latter to be an interesting question, too. But it is not the question I mean to be discussing.

WTD could arise in the case of a person who thinks he knows the truth about the normative ethics of behavior. Take me, for example, I think some form of desert adjusted hedonic act utilitarianism is true. Yet in virtually every interesting real-life case I am sure I don't know the relevant utilities of any of my alternatives. Indeed, I generally am pretty much in the dark about what my alternatives are. Thus, even though I think I know in abstract

terms what I should do ("whatever would maximize desert adjusted utility") it often happens that I don't know for sure what
my alternatives are, and I have no idea which of my alternatives
is the one that satisfies this criterion. Thus, in a real practical
sense, I don't know what I should do. This is a question that has
confronted me on many occasions. It sometimes seems to me that
this question arises on virtually *every interesting occasion of moral
decision.*

Other people are even worse off. Like me, they don't know their
alternatives. But in addition they may not have any firm convictions about moral theory. They don't know what would make one
of their alternatives morally right. Still, such a person might be
interested in doing the right thing.

Some advocates of utilitarian theories have suggested that when
you don't know what is actually right (= what maximizes utility)
you should perform whichever of your alternatives maximizes expected utility. As soon as we spell this out in precise detail, its
absurdity becomes apparent. To determine which of your alternatives maximizes expected utility, you need to proceed as follows:
first, list your alternatives. Second, select the first alternative, list
all of its possible outcomes. Third, for each possible outcome, list
its value and its probability given the action. Fourth, for each
outcome multiply probability times value. Fifth, find the sum of
the products for all outcomes of the first alternative. The result is
the expected utility of the first alternative. Repeat this procedure
for each alternative. Then identify the alternative that maximizes
expected utility. Perform that alternative.

The question was: 'what should I do when I don't know what
I should do?' The proposed answer, suggested by some utilitarians, is that you should do whatever maximizes expected utility.
It should be obvious, however, that if a person does not know
what maximizes regular utility, he will almost certainly be in no
position to determine what maximizes expected utility. The latter
computations are dramatically more complex, and they rely on
dramatically greater funds of information about the values and
probabilities of possible outcomes of alternatives.[14]

Furthermore, if the utilitarian assumes that what you should do
is whatever maximizes actual utility, then it immediately follows

---

[14]I developed this line of argument at much greater length in my 'Actual
Utility, the Objection from Impracticality, and the Move to Expected Utility',
*Philosophical Studies*, 129, 1 (May, 2006): 49-79.

that you should do the thing that maximizes expected utility *only if that same act also happens to maximize actual utility*. Suppose (as is obviously possible) that some act, a1, maximizes actual utility. Suppose some alternative, a2, does not maximize actual utility but does maximize expected utility. Then it is simply wrong for a utilitarian to say that if you don't know what you should do (which is, according to his view, a1) you should do what maximizes expected utility (which is, in this case, a2). His own theory entails that you should *not* do a2.

G.E. Moore once seemed to suggest that when you don't know what you ought to do, you ought to do whatever is required by conventional morality.[15] This also seems hopeless. In the first place, if what is required by conventional morality is different from what is required by the true moral theory (Moore would presumably say that this is the act that maximizes ideal utility) then the Moorean advice is simply wrong – nearly self contradictory. For suppose you have two choices and the first is actually obligatory but you don't know this. Suppose conventional morality says you should perform the second choice. Then the claim that you ought to do what is required by conventional morality is simply false. *Ex hypothesi*, you ought to perform the first choice, and conventional morality says you ought to do something else.

Another problem with Moore's suggestion is that it is often difficult to determine what conventional morality requires. In many controversial cases, there is debate about what conventional morality requires. In other cases, with inadequate information about circumstances and consequences, even a person who knows a lot about conventional morality would be hard pressed to say what it implies for his particular circumstances.

Others have drawn a distinction between "objective obligation" and "subjective obligation", where the former is determined by the true moral theory together with the actual facts of one's circumstances and the latter is determined by one's generally sketchy and incomplete evidence concerning theory and circumstances. The proposed answer to WTD is then: When you don't know what you objectively ought to do, you *ought* to do whatever you subjectively ought to do.

---

[15] I am alluding to the notorious Section 99 of Chapter V of *Principia Ethica* where Moore says, among other things, 'In short, though we may be sure that there are cases where the rule [of commonsense morality] should be broken, we can never know which those cases are, and ought, therefore, never to break it.'

I have a question about the italicized 'ought'. If it means 'objectively ought' then the answer is clearly wrong. It is about as plausible as this statement about mountains: If you don't know that Mt. Everest is the tallest mountain, then the tallest mountain is whatever mountain you think is tallest. Just as your ignorance of geography cannot affect the heights of mountains, so your ignorance of your obligations cannot make something that is not your objective obligation become your objective obligation. On the other hand, if the underlined 'ought' means 'subjectively ought' then the statement as a whole is completely trivial. For it ends up saying that when you don't know what you objectively ought to do, then your subjective obligation is your subjective obligation. But of course your subjective obligation is always your subjective obligation. How could pointing out this triviality be helpful to a person who wants to know what he objectively ought to do?

It seems to me that this question about the interaction between inadequate information and moral obligation is deep and important. It also seems to me that it is one of the questions we most frequently get from people who are not moral philosophers. For several years I served as chair of the Ethics Advisory Board of a biotechnology firm. People associated with that firm often seemed to be concerned about the morality of various proposed lines of research. They sometimes turned to me, asking me for advice (which seemed pretty reasonable, given my title). Yet even though I was fairly well convinced of my moral theory, I was never confident of any answer to the questions posed. Since I was always in doubt about the consequences of various alternatives (would the proposed research lead to a cure for cancer? or would it lead to illness and death among the researchers?) I was never able to offer much useful guidance. Others who share my interest in normative ethics, if they are honest about it, will have to admit that they also find frequently themselves in the same situation.

Thus it seems to me that one of the most important and overlooked questions confronting normative ethics is WTD: what should I do when I don't know what I should do? Although I am pessimistic about my own chances of finding a good answer, I hope that others will become convinced that it is a worthy question. Perhaps someone with greater insight will decide to devote some thought to this question. With luck, perhaps he or she will come up with a plausible answer.

# 5

# David Heyd

## Chaim Perelman Professor of Philosophy
The Hebrew University of Jerusalem, Israel

---

> Meeting a friend in a corridor, Wittgenstein said: 'Tell
> me, why do people always say it was *natural* for men to
> assume that the sun went round the earth rather than
> that the earth was rotating?' His friend said, 'Well,
> obviously, because it just *looks* as if the sun is going
> round the earth.' To which the philosopher replied,
> 'Well, what would it have looked like if it had looked
> as if the earth was rotating?'
>
> —Tom Stoppard, *Jumpers*, Act II

Philosophy exerts two opposite "pulls" on anyone engaged in it:
the Platonic and the Socratic. The first instills in the philosopher
the pretentious ambition to uncover truth and give meaning to
the world, and do so in a direct and systematic way (unlike art)
and within a comprehensive and universal scope (unlike science).
From the optimistic point of view informing this approach, the
world looks both transparent and "accountable", i.e. susceptible
to rational explanation. Philosophical effort is expected to yield a
theory (traditionally called a "system"), or at least theories. The
other driving force of philosophical reflection is more humble and
modest, constantly sensitive to the limits of what we may hope
to know. It advocates caution, suspicion and self-awareness of our
inherent fallibility. What philosophy can (and should) add to our
conception of the world is critique rather than theory, skepticism
rather than doctrine.

This duality of opposite intellectual thrusts is not surprising
since it reflects the inherent structure of human consciousness itself. This irreducible and unique faculty of human beings is Janus-
faced: it propels us to transcend the bounds of our immediate experience and look at ourselves from the outside, as part of the

world; but at the same time, being self-aware, human consciousness views such transcendence as illusory, a mere construction.
Consciousness shows the way out of itself but is at the same time
imprisoned in its own bounds. Although these two impulsions of
philosophy have accompanied it from its inception, highlighting
this problematic duality as its fundamental feature is a typically
modern phenomenon. Kant's philosophy moves back and forth
from the dialectic of reason to its critique. Tom Nagel's work lies
on the axis of the "view from nowhere" and the inescapably subjective perspective. Both philosophers believe that this philosophical tension characterizes morality no less than epistemology and
metaphysics.

Thus, much under the influence of philosophers like Kant and
Nagel, philosophy has always seemed to me to be an enterprise
of "bootstrapping", an effort to extend human vision so as to
include the eye itself. Despite the apparent paradoxical nature of
this project, it is what gives philosophy both its clarity and depth.
In retrospect, it does not surprise me that I was attracted to what
might be the most concise and pure case in moral philosophy
illustrating this tension between the subjective (internal) and objective (external) perspectives, namely the morality of procreation
or what I have referred to as "genethics". This mind-boggling subject has become one of the main foci of my work in the past two
decades.

Few philosophers are fortunate (or rather creative) enough to
be remembered as having started a *new* field of discussion. In
his Oxford seminars in the 1970's and then in his seminal book
*Reasons and Persons*, Derek Parfit has rightly earned that title.
Taking part in one of those seminars as a Ph. D. student and
then reading the book in the beginning of my professional career,
I was struck by the theoretical depth of Parfit's question: what
sort of moral principles can guide the decision to bring people
into the world, how many of them, and under what identity? Although looking like any other question in ethics, concerned with
the proper way to deal with other human beings, these questions
force us to articulate the general conceptions of value and valuation which underlie all our ethical discourse.

How is this? Consider what might be the purest and simplest
illustration of "genethical" dilemmas, the so-called "wrongful life"
cases. With what I learnt from Parfit in the back of my mind,
I came across a controversial decision in 1986 by the Supreme
Court of Israel to grant legal standing to a child who sued the

genetic counselor of his mother for having led to his birth in defect. The mother, who had a brother suffering from Hunter's disease, approached the counselor with the question about the risk of her future child of suffering from that very serious and incapacitating disease. She was determined not to conceive a child if there was a substantial genetic risk involved. The counselor, out of negligence, assured the woman that there was no cause for concern and on that basis the woman went on to conceive a child who was born afflicted with that disease. Now, it was obvious that the parents could on their part sue the counselor for damages, but could the *child* do so? Can a child be harmed by being given life (even if this life is of low quality)? Can it be wronged by being born? Can people have a right not to be born? Or, a right not to be born in defect? And if parents (or genetic counselors) have a duty (of care? but care for whom?) to prevent the conception of severely handicapped children, do they have a similar duty to produce healthy and happy children?

The way we deal with these questions turns primarily on the role of identity in moral (and legal) evaluation. For the main argument for denying the child's standing in wrongful life cases is that had the negligent act not taken place, *he* would not have existed, and hence he cannot claim that *his* condition worsened as the result of the act. This powerful challenge, referred to since Parfit as "the non-identity problem", forces us to choose between two fundamental conceptions of moral evaluation: the person-affecting and the impersonal. According to the former, values are always "good for" people (persons), i.e. they concern the way people are *affected* for better or for worse in terms of their welfare, satisfaction of their needs, promotion of their interests, fulfillment of their ideals, etc. According to the latter, values are attributed to "the world" in general, making it better or worse independently of the experiences, interests and needs of actual people. In most moral judgments the distinction is of little theoretical import, since benefiting and harming people are coextensive with benefiting and harming *tout court*. However, there is one exception and that is the creation of new people—adding happy (or miserable) individuals to the world. According to the impersonal approach, such additions are good (or bad, respectively). According to the person-affecting approach, these additions cannot be judged as morally good (or bad), since there are no actual persons who are affected by the additions for better or for worse (assuming that the existing people are indifferent to these additions). For both being born and not being born cannot be considered from that point of view as either a benefit or a harm.

Having developed in my book *Genethics* a defense of a strongly person-affecting conception, I have become skeptical of our ability to make moral judgments about procreation, genetic selection, population policies and even certain forms of education (those which decide "identity fixing" characteristics of children). From a strictly anthropocentric perspective, all we have is the given fact of the existence of human beings (in general, and the actual ones in particular). The discourse of valuation takes off the ground only once human beings exist, both as the subjects making those valuations and as objects of these valuations. The debate between the two camps regarding the nature of value is deep and it seems there is no knock-down argument for either of these views and that the adoption of either of them may be defended only by global considerations regarding the more coherent manner in which we can reconcile our intuitions in *various* contexts with the theoretical conception. But the issue remains a clear example of the bootstrapping nature of philosophical reflection. For even if we believe (as I do) that values (including moral values), duties and rights apply only to actual human beings, there is an irresistible "pull" to examine, as if from the outside, the value of the very existence of human beings. Beyond the example of wrongful life, an illuminating thought-experiment is the possibility of the total extinction of the human species, or to make it even harder for our imagination, the possibility that the human species did not evolve at all. Can it be said to be good or bad? To put it as a paradoxical question: is there value in the existence of the preconditions of value?

The problem is structurally analogous to Kant's: can we say anything about the validity of our rational judgments about the world in terms which are not constructed by reason itself? Can we rid ourselves completely (as we are required by the critical analysis of reason) from the external point of view from which we can allegedly judge the place of reason in the world "itself"? As Kant has taught us in the *Critique of Judgment* and in the end of the *Critique of Pure Reason*, the aspiration to transcend the bounds of human judgment (or concepts) is unstoppable, although it must always be checked and restrained. The ethics of procreation is again the sharpest example in ethics for this irresolvable tension. But I have been attracted also to other manifestations of this problem, such as the theory of justice. For the Greeks, justice was an essential property of the natural world and of the structure of the human soul (as well as that of society). But for modern conceptions, such as John Rawls', justice is a human ("Kantian") *construction*, that is to say a product of human reasoning and consent rather than of the discovery of a given natural relationship. However, if that is the case, the scope of application of the principles of justice is limited to actual (given) human beings and the principles of justice cannot govern our "genethical" choices, i.e. how many people to produce and in what mold or identity should they be shaped. This again makes the question of inter-generational justice not only a pressing moral and political issue, but a deep theoretical stumbling block.

Rawls, in his later work, reached the conclusion that the only duty of justice we have to future generations is the sustenance of the conditions of a just society. But from the constructivist or person-affecting point of view, it seems a bootstrapping feat to argue that it is just to maintain justice! It is impossible to apply principles of justice to the existence of the conditions of justice. Or to put it in the terminology used by Hume and Rawls, the existence of the "circumstances of justice" (such as the ability of human beings cooperate or their limited altruism) is not itself either just or unjust. In my work on intergenerational justice I have accordingly reached the conclusion that the only way to explain the duties of justice to future people lies in our aspiration to self-transcendence, i.e. to the fulfillment of *our* life projects in periods extending beyond our own life span. We choose to shape the structure of future society not in the light of the rights or interests of future people but rather in ways which would represent our long-term aims and reflect the values that make our lives meaningful to us. Justice is not an "impersonal" value which we are obliged to promote as such; nor is it a guiding principle for making demographic choices; it is rather a person-affecting value which we believe is good for us and which we want our children to believe is good for them too.

*** 

My formative years as a philosophy student in the early 1960's coincided with the gradual shift of focus from the meta-ethical to the normative approach in moral philosophy. The battle between prescriptivism and naturalism was raging between Richard Hare and Philippa Foot and everybody was engaged with how to derive, or not to derive, "ought" from "is". A common first-year exercise was to destroy emotivism (which was easier) and intuitionism (which was harder). But Rawls' early work on justice as fairness has already captured our minds. And there was also a 1958 short paper by J.O. Urmson, called "On Saints and Heroes", which again had that rare fortune of having single-handedly revived a subject that has been dormant for centuries, that of supererogation. Besides his extensive work in meta-ethics, Urmson contributed to normative theory by breaking the rigid tripartite classification of actions into the permissible, the obligatory and the prohibited, a classification that was shared by both deontologists and utilitarians, and adding a fourth category of saintly and heroic acts.

I chose supererogation as the subject of my Oxford dissertation, expanding on Urmson's discussion, especially by anchoring it in the long, rich and sophisticated historical debates in philosophy and theology about the possibility and desirability of acting beyond the call of duty. In my book *Supererogation* I tried to show the general theoretical contribution of the concept to the normative analysis of the relationship between the right and the good and the failure of both Kantian and utilitarian ethics to account for what might seem a small part of our daily behavior but one which has immense moral value. My own work, and that of many others in the past three decades, has shown the crucial role this concept plays in the analysis not only of heroic and saintly acts of the kind Urmson was speaking of but of forgiveness, gifts, and even (as I have lately tried to argue) toleration. The genuine and pure act of giving or forgiving is completely free, that is to say, it is not even a weakened kind of duty, an ideal duty or a duty directed at those who have an especially strong moral character. The supererogatory is an irreducible moral category, without which our moral life would have been significantly impoverished.

Why is it so? Again, as in "genethical" issues, the problem of supererogation touches upon one of the fundamental crossroads in ethical theory: the relationship between the act and the actor. One of the weaknesses of deontological and consequentialist theories of the $20^{th}$ century was their exclusive focus on the act. Elizabeth Anscombe and Bernard Williams (among others) have turned our attention back to the classical emphasis on the agent—the one by reviving the idea of virtue ethics, the other by highlighting the importance of moral integrity and the role of character in moral judgment. Now, it seems to me that the special value of supererogatory behavior has to do with the particular intention of the agent and not necessarily with the beneficial outcome of his act, for it is often the case (as in small gifts, charitable forgiveness, or the tolerant restraint from insisting on one's rights) that the outcome is relatively insignificant. Supererogatory action, in all its manifestations, promotes personal relations, that is adds a personal dimension to the impersonal system of justice, rights and duties. I have also argued that the split between action and agent may serve to solve the famous paradox of toleration, namely how can intentional restraint from interfering in another's (objectively) wrong behavior be considered virtuous.

Another topic in which this shift from act to agent is manifest is that of moral luck. It could be nice if we, the contributors to this volume, could answer the question what are the most neglected topics in normative ethics. But we can at least point to those original and imaginative minds of our era, like Bernard Williams, who had the insight and foresight to do so. Here is the third example of a single article starting a new field of philosophical discussion, which by now has yielded a vast body of literature. Williams' 1979 paper "Moral Luck" challenged not only our traditional, over-legalized notions of responsibility but also the very possibility of moral theory, calling for a re-mapping of the contours of "ethics". The basic idea is that human life is inescapably susceptible to conditions over which we have no control, circumstances that are from our point of view a matter of luck. Even those acts (most of our acts?) which are not strictly under our control and choice are *ours*, that is to say part of what we are and do. Human agency is saturated with contingency and moral theory cannot idealize and purify it in the way Kant thought was possible. Even if we are not strictly responsible (or blameworthy) for running over a pedestrian who recklessly bumps into our car, we do and should feel what Williams called "*agent* regret". Even if an anti-Nazi citizen of Germany in the 1940's should not be held responsible for the deeds of his government, he should feel *shame* for being German (as the German philosopher, Karl Jaspers has argued). Shame, unlike guilt, applies to what I am rather than to what I did.

***

Moral philosophy is in its nature normative, that is to say strives to guide human behavior. But being a philosophical enterprise, it focuses on general principles, on idealized conditions, on the ideals and utopias which are free from the constraints of contingent realities—political or institutional. An unprecedented revolution in ethics has occurred in the second half of the twentieth

century with the surge of "applied ethics", a trend which started
with medical ethics but rapidly spread to other departments of
professional ethics (business, media, military, and many others).
The main threat to moral philosophy that I detect in this development is *fragmentation*, the dissection of human behavior into
narrow professional contexts, each having its own rules, values
and duties. This attempt to interpret the normative role of ethics
in a directly "regulative" manner can undermine the traditional
enterprise of moral philosophy, which aimed at abstract principles
connecting values and principles of action with the general traits
of human nature. The proliferation of sub-fields in ethics runs the
risk of transforming moral philosophy into an amalgam of ad hoc
manuals for the proper functioning of human beings *as* members
of this or that profession (a phenomenon most typically manifest
in the constantly growing number of "ethical codes"). Culturally
speaking, this tendency to the applied aspects of ethics is an expression of philosophy's attempt to prove its "relevance" in modern liberal societies, an effort which is partly a genuine response
to novel moral dilemmas and partly apologetic in nature. While
moral philosophy in the mid-twentieth century was confined to
self-contained, esoteric theoretical analysis of the language and
logic of moral discourse, with the intentional avoidance of making
any judgments on substantive moral issues, contemporary moral
philosophy is often too keen to take sides in public debates on
urgent matters of everyday life. Against the "linguistic turn" of
ethics, beginning with G.E. Moore, a sharp "normative turn" took
place in that field in the past four decades, much under the influence of the rise of political philosophy and John Rawls' work on
justice. But with applied ethics, philosophy, which was considered
as having nothing to say about ordinary life, is now often thought
as being able to say something on every moral issue in our public
life.

It cannot be denied, however, that bioethics (as arguably the
primary example of applied ethics) has had an important impact
on the development of ethical thinking. The debates about abortion and euthanasia led to a deeper look at the meaning and value
of human life. The discussion of priorities in health care gave rise
to sophisticated considerations in the theory of distributive justice. Defining the conditions of informed consent has highlighted
the epistemological and volitional factors in the general theory
of consent. The controversy over stem cell research and prena-
tal genetic diagnosis triggered the extension of the discussions of

"wrongful life" to those of "wrongful identity". And the list can go on and of course extend to other fields in applied ethics. On the other hand, it is hard to conceive of any serious contribution to applied ethics which would not be heavily informed by general moral theory that is based on conceptual analysis, abstract principles and a metaphysical view of human beings. Be it due to internal changes in the way moral philosophers regard their role or to the external expectation of the public that philosophers provide guidance in resolving new dilemmas in modern life – the extension of moral philosophy to the applied and professional sub-fields seems to be irreversible. Hence, one of the main challenges of normative ethics in the future will be to adapt to this change without losing the more traditional characteristics of moral philosophy.

This is a difficult challenge since the two approaches to moral issues may often be hard to reconcile, as I found out in my personal experience as a player on both fields. Ethical theorists and applied moral philosophers are guided by different commitments: the former, to theoretical coherence and consistency; the latter, to a balanced response to social concerns and sensitivities. Theory is cost-free; regulation is a socially risky business. No wonder that ethics committees (like court decisions) tend to be more conservative than some of the results of abstract philosophizing. Having read my article about wrongful life, one of the judges in the Israeli court decision told me that although my theoretical argument for denying legal standing from the child made a lot of sense (since had the negligent act not taken place *this* child would not have been born at all), he, as a judge, was forced to take into account social, economic and humanitarian considerations in awarding "compensation" to the child. Or, having heard me argue for a lottery system in some cases of the allocation of scarce resources in organ transplantations, a chief cardiac surgeon told me that although he shared my view, he could never declare this to be his guiding principle due to the risk of losing public credibility. The chasm between two approaches is real. And it is difficult to bridge also due to the fact that many of the people engaged in applied ethics are not philosophers in their training. But beyond this professional or disciplinary split, it is definitely important to develop a *philosophical* discussion of the role of applied ethics in moral philosophy. After all, it goes down to the very identity of "normative ethics"

What are the most important problems in normative ethics is of course a subjective matter. But taking my cue from the preceding description of the uneasy but potentially fertile association of the theoretical and the "applied", I can guess that new possibilities in the genetic shaping of future generations will present a major – even unprecedented – challenge to moral philosophy in the near future. This may prove to be the meeting grounds between the ethics of genetics (in the applied sense) and "genethics" (in the theoretical sense that I have defined). Scientific and technological developments in genetics will impose new dilemmas on society and individuals: What is human nature? Is there a moral commitment to its preservation and perpetuation? Is there anything morally wrong in creating sub-human or super-human beings who will not have those properties which make *us* moral subjects? These are not just new moral issues like prioritization in health care or the limits of informed consent, since they relate not to the rights and duties towards human subjects (persons) but to the molding of those subjects into some non- or quasi-human form (clones, hybrids, biological automatons, pre-embryonic suppliers of stem cells used for therapy).

This complex of questions turns us back to the bootstrapping character of the philosophical enterprise with which we started. For is it not exactly what we are trying to do by extracting ourselves from the position of being subjected to the laws of evolution to becoming its masters? And does not the attempt to evaluate the transformation of our essential human identity require the adoption of some external point of view, transcending the human one? Jürgen Habermas' *The Future of Human Nature* is an impressive (even if not always convincing) attempt at such self-transcendence. But guessing or even suggesting the important problems of normative ethics for the future does not imply a belief in the "prospects for progress". Looking back at the long and rich history of normative ethics, I cannot trace any trend of progress in the normative sense of "becoming better", and hence cannot project progress to the future of the field. In that respect, I am not sure that Parfit's belief or hope that in the future we will *discover* "Theory X", which will solve the paradoxes of non-identity by combining the person-affecting and impersonal approaches, is realistic. The riddle of the morality of procreation will not be solved. It will rather be resolved when our notions of personal and species identity will gain new, unpredictable form.

# 6

# Thomas Hurka

## Jackman Distinguished Chair
## in Philosophical Studies
## Fellow of the Royal Society of Canada
University of Toronto, Canada

I became interested in normative ethics in my last term as a philosophy undergraduate at the University of Toronto. Influenced by a traditional conception of the discipline, I'd till then studied mostly history of philosophy, with a special interest in, of all things, Hegel. But seeing the value of a balanced philosophy program, I enrolled in an ethics seminar in the winter of 1975. I'd studied the ethics of Plato, Leibniz, Hegel, and others in my history courses, but this was my first exposure to contemporary thought on the subject.

The seminar had only one other student, and since he was fairly quiet I had something close to a one-on-one tutorial with the instructor, Wayne Sumner – another contributor to this volume. I was immensely taken both by Wayne's teaching and by the seminar's topic, which was utilitarianism. Our study of it included both theoretical questions such as act- vs. rule-utilitarianism and applied ones such as our obligations to future generations. I found all these questions intellectually engaging and also humanly important, given their connection to live ethical issues – in short, I was hooked.

I was especially impressed by two non-utilitarian texts we read. One was the last chapter of G.E. Moore's *Principia Ethica*, meant to illustrate a non-utilitarian or ideal consequentialist theory. I was attracted by Moore's general account of the good, which values states such as love and aesthetic appreciation alongside pleasure, and saw in his principle of organic unities a far clearer statement of ideas I had encountered in more pretentious form in Hegel. The second was Chapter 2 of W.D. Ross's *The Right and the Good*, which defends a pluralist deontology against consequentialism. I thought Ross's criticisms of consequentialism were right on the mark, and again clearly and precisely expressed. I had no idea these texts were largely *passé* in current discussions of the subject.

At this time I was applying to graduate programs in philosophy, and given my new interest decided to specialize in ethics. I chose Oxford as my university, and when I got there, spent some time studying metaethics, for example, emotivism and prescriptivism, with R.M. Hare. But the main activity in Oxford then was in normative ethics. There were joint seminars on normative topics each led by (some among) Ronald Dworkin, James Griffin, Jonathan Glover, J.L. Mackie, Derek Parfit, Amartya Sen, and Charles Taylor, as well as seminars by individual philosophers, such as Parfit's on what would become *Reasons and Persons*. They all used the same analytic methodology I'd been impressed by in the Toronto seminar, though on a wider range of topics. And I wanted to use it on a wider range still.

My thesis idea was to use this methodology to defend an ethical view that I initially called 'self-realizationist' and later 'perfectionist.' Understood broadly, this view tells us to promote a conception of the good that, like Moore's, includes 'ideal' states such as knowledge and achievement alongside or instead of pleasure or desire-fulfilment. Understood more narrowly, it identifies these ideal goods as developing properties that are somehow fundamental to human nature, for example, by being essential to humans. This view wasn't being discussed much in Oxford or indeed anywhere in analytic philosophy, but it was one I'd encountered and been attracted to while studying Aristotle, Leibniz, Hegel, and other historical figures at Toronto and that I thought could be rehabilitated using ideas like Saul Kripke's account of essential properties. The idea was to put old perfectionist wine in new bottles.

It's said that the key to a successful thesis is to pick a 'sunrise' rather than a 'sunset' topic, one that other people have started to write on but that hasn't yet been worked to death. In this terminology mine was a pre-dawn topic, and in working on it I was largely on my own. This led to some floundering; for example, it took me a long time to see that perfectionism can tell each person to promote the development of everyone's human nature rather than just her own. I worked away at the topic through a B. Phil. and D. Phil. thesis to, finally, in 1993, a published book called *Perfectionism*.

Over time my treatments became, I hope, less inadequate, but they also changed focus. The B. Phil. thesis was almost entirely devoted to the narrow perfectionist idea of grounding the good in human nature, with very little on the details of the moral view that would result. The D. Phil. was more evenly balanced between the two topics, while the book gave the bulk of its attention to the detailed theory. After three chapters on human nature, it had eight about a broadly perfectionist theory that could be based on ideas about human nature but could also be defended apart from them; these chapters discussed, among other things, the aggregation and balancing of perfectionist goods, the parallels between knowledge and achievement as goods, and the political implications of a theory centred on them.

In the years since, my thinking has moved further in that direction, so I'm now more sceptical about grounding the good in human nature. It's not that I don't think the narrow perfectionist idea has intuitive appeal; I still think it does. But I'm less confident that a dispassionate examination of our best explanations of human behaviour really does pick out all and only those properties that are intrinsically worth developing. I'm more inclined to think that knowledge and achievement – *Perfectionism*'s account of which I still stand by – are best valued on their own rather than as based in human nature, as are other states such as virtue that can't be accommodated in the narrow perfectionist framework.

And my shift in attitude has been more general. Normative ethics tries to systematize our everyday moral judgements by relating them to principles that are somehow more explanatory. But there are two different ways of doing this, or two ends on a continuum of ways.

One relates everyday judgements to principles that are more abstract but at the same time continuous with those judgements, because they use similar concepts and concern a similar topic. An example is Moore's account of retributive desert using his principle of organic unities: while vice and pain are both evil, he says, the combination of vice and pain in the same life is good as a combination, and sufficiently so that inflicting the pain is on balance an improvement. This account reveals the underlying structure of retributive judgements, connects them to others involving organic unities, and suggests further questions, such as how the goodness of deserved pain compares with the evil of vice. But it doesn't justify the judgements in other terms and so won't persuade anyone not already sympathetic to retributivism. Another example is the claim that we have stronger duties to promote the happiness of our children, friends, and other intimates because we have in general stronger duties concerning people who stand in certain special relations to us. This again reveals the structure of associative duties and suggests further questions, such as what exactly the relevant relations are, but it doesn't justify the duties in non-associative terms. I call this style of normative theorizing *structural*.

Many contemporary philosophers find this approach unsatisfactory. They say structural principles are too close to the everyday judgements they're meant to explain: they in effect assume what they're meant to justify and so don't justify at all. That requires relating a moral judgement to claims that use different concepts and concern some other, more fundamental topic, one that does yield justifications. The resulting *foundational* style of theorizing has several variants. Some of these try to ground moral judgements outside morality itself, for example in scientific theories about psychology or evolutionary biology, in metaphysical theories of the self, or in semantic claims about the meanings of the moral words. Others relate everyday judgements to claims within morality, but on some other, allegedly more fundamental topic. Thus, contemporary neo-Aristotelians say the reason we should promote others' happiness is that this will express benevolence on our part, where virtues like benevolence are essential for our flourishing, which of necessity is our ultimate aim. Another exemplar of this approach is John Rawls's *A Theory of Justice*, which assumes that claims about liberty and equality can't be affirmed just on their own. They're properly justified only if we can show that they would be chosen by rational contractors in what Rawls calls the 'original position.' This position isn't independent of morality, since we use moral judgements in specifying its features. But it's far removed from everyday thinking about political values, which makes no reference to rational contracting.

The narrow perfectionism discussed in my book was to a considerable extent foundationalist. Unlike many versions of similar ideas, it didn't try to ground the human good directly in non-evaluative claims about human nature; instead, it affirmed a substantive moral principle valuing whatever properties that nature turns out to involve. But ideas about human nature are fairly distant from everyday thought about the value of understanding or achievement, and I now extend my scepticism about those ideas to foundational theorizing in general. However impressive it may seem, and however exciting the justifications it promises to deliver, it rarely if ever succeeds.

This can't be shown across the board; individual foundationalist arguments must be examined individually. But time and again they turn out not to yield the conclusions they're meant to, or not without tacitly assuming what they're meant to prove. (The latter was the foundationalist charge against structural analyses, but they don't claim to explain on a wholly different basis.) Consider the neo-Aristotelian idea that the virtues are those traits a person needs in order to flourish or live well. To yield the intended kind of explanation, it needs an independent understanding of what human flourishing consists in, which many give in terms of rationality. But if rationality is understood in a not morally loaded way, it can surely be exercised in malicious acts as much as in benevolent ones. And if it's instead taken to include a power to know and be guided by true practical principles such as 'Promote others' happiness,' then it assumes its intended conclusion.

And there's a further problem: many foundational arguments distort the phenomena at issue by turning the duties everyday thought recognizes into something they're not. Consider again the neo-Aristotelian view. To the question, 'What's the ultimate reason why you should promote another's happiness?' it answers 'Because that will make your own life better,' and that's not the right answer. The right answer is 'Because it will make the other person's life better.' Here the foundational argument has turned an other-regarding duty into an egoistic one, and the same occurs in Rawls's theory. He thinks fairness requires his contractors not to know their particular conception of the good, or the specific values they're pursuing. But he then assumes that they'll try to maximize the resources they'll need to pursue whatever conception of the good they have. And that makes their primary concern to live, not a life that's in fact good, but whatever life they think is good, so they care about their conception of the good not because it's true but because it's theirs.

This leaves the alternative, structural style of theorizing, and while it doesn't yield as grand explanations as foundationalism, it does yield explanations. It connects everyday judgements to principles that are more general and therefore more explanatory and that are also often intuitively appealing in themselves, or apart from their implications. To illustrate this, consider the very different, non-Aristotelian account of virtue discussed in my second book, *Virtue, Vice, and Value.*

This account takes virtue to consist in morally fitting attitudes to other, previously given goods and evils. If something is intrinsically good, then loving it for itself, or desiring, pursuing, and taking pleasure in it for itself, is also intrinsically good and a virtue. So if another person's pleasure is good, then desiring, pursuing, and taking pleasure in his pleasure is good and, more specifically, benevolent. Similarly, if another person's pain is evil, then hating his pain, or desiring its absence and being pained by its presence, is intrinsically good and compassionate. But loving another's evil is the evil of malice, while hating his good is envy.

This account remains reasonably close to everyday thought, which would agree that benevolence involves caring positively about what's good for other people and compassion being against their evil. But it expresses these ideas in a more abstract and therefore explanatory form. If we have the general idea of loving the good, we can collect under it not only the virtue of benevolently desiring another's pleasure but also those of disinterestedly pursuing knowledge, if knowledge is good, desiring just distributions, and more. Going further, we can unify all the virtues under the general heading of attitudes appropriate to their objects, because their orientation matches their objects' value, either positive to positive or negative to negative. This type of unification is explanatory, especially when the abstract principles are intuitively appealing in themselves, as I believe they are. It's attractive just as an abstract idea that loving a good is another good and loving an evil an evil.

At the same time, the account suggests further questions of detail. If attitudes can be more or less virtuous, how does their degree of value depend on their intensity and the degree of value of their object? What about attitudes to evaluatively neutral objects, or indifference to good or bad ones? And how do the different forms of virtuous love, such as desire for absent goods and pleasure in present ones, compare with each other? If we can answer these questions, and in a systematically connected way, that will again deepen our understanding.

This account of virtue is by no means my invention. It was widely accepted among moral philosophers of the late $19^{th}$ and early $20^{th}$ centuries, among them Moore and Ross. And in fact this period's philosophers consistently favoured structural over foundational theorizing. They rejected the more ambitious approach of ancient ethicists such as Aristotle, in part because it was obnoxiously egoistic, and likewise rejected Kant's attempt to derive specific duties from the bare idea of consistent rational willing. I think the closeness of Moore's and Ross's analyses to everyday moral thought, and their resulting clear relevance to that thought, is part of what attracted me to them as an undergraduate and sustained my admiration for them even when I knew they were out of philosophical fashion. In now more explicitly favouring structural analysis I'm returning to that initial attraction.

In recent years there's been much valuable writing in the structural style. In the theory of value there have been more elaborate analyses of equality, desert, and other goods than any given by Moore or Ross, and there have been similar advances in understanding deontological ethics. Though Ross held that there's a more stringent duty not to harm other people than to benefit them, he said little about what the difference between the two consists in. The recent voluminous literature on the doing/allowing, intending/foreseeing, and other distinctions, though not conclusive, has greatly illuminated the philosophical options in this area. It's likewise been emphasized that deontological ethics is structurally agent-relative, so it doesn't tell people to minimize the total number of, say, lies told, but to care specially that *they* don't lie. And despite his effort to remain true to common-sense morality, Ross held that, when other duties are silent, we're required to maximize the good impartially. Against this, more recent writers have shown that everyday thought is less demanding, allowing agents to care somewhat more about their own good than about other people's and therefore sometimes not to do what's impartially best.

At the same time, however, immense energy has in recent decades gone into foundational projects, which probably have the greatest prestige among philosophers. They include, alongside neo-Aristotelian and Rawlsian views, a grandiose Kantian argument that commitment to specific moral principles is implicit in the presuppositions of all rational agency. Even some of the structural discoveries mentioned above have generated foundational demands: philosophers have required the agent-relativity of deontological duties to be given some separate and deeper justification, or have tried to ground the permission to favour one's own good in metaphysical claims about the self and its points of view.

If, as I have argued, foundational theorizing rarely succeeds, these efforts are unlikely to yield positive results. But they also distract philosophers from what I think would be more fruitful work. Someone who's seeking the deep foundation of morality is less likely to look at the details of what morality says; her attention is somewhere else. And there's a more insidious effect. The more complex morality is, and the more subtle its features, the harder it will be to justify on conceptually separate grounds, just because the justification has to justify more. There's therefore a temptation for foundational theorists to simplify the content of morality, so their justifications have less to do. (The alternative is to leave the details to 'true practical principles' or 'what the virtuous person perceives,' but that's simply to give up on explaining what can be explained by structural means.) In both ways, more interest in grand foundational theorizing leads to less interest in the details of our actual moral scheme.

But those details are fascinating, surprising, and eminently worth study. Our everyday intuitions often make fine-grained distinctions, about either value or duty, whose underlying rationale it's illuminating to identify. And those intuitions can have more integrity than appears: sometimes an attractive moral claim turns out to entail others on the same topic that initially seemed independent of it. Or there can be unexpected connections between topics, with our judgements about one having the same structure as about another, so the two sets of judgements run in parallel. In all these cases structural analysis can show everyday thought to be more coherent than first appears.

Of course we may not always find this coherence. For some fine-grained judgements, for example about when one may cause harm to prevent greater harm to others, a systematizing principle has proved very hard to formulate. And if in fact there's no such principle, that may be grounds to question the judgements' reliability. Or if there is a principle but it has no independent appeal – its only merit is to yield the right particular judgements – that too may be grounds for doubt. And of course we will often encounter disagreements about particular judgements. Everyday moral thought isn't a monolith. It contains different elements, which different people emphasize to different degrees, leading to disputes about what's right and wrong in particular cases. So everyday thought contains not just one view but a plurality of partly competing ones. In all these cases, however, structural analysis remains the best response. If our everyday judgements can't be systematized, a serious effort to do so is the only way to establish that fact. And when people disagree, a similar analysis can pinpoint the exact crux of their dispute, increasing the possibility of resolution and at least making clear what the ground of difference is.

It may be objected that structural theory is inherently conservative, confined to systematizing existing moral beliefs rather than changing them in fundamental ways. But foundational work is also often conservative, either in practice or also, as in Rawls, in its explicit intent. And the structural approach is no less able to play a reforming role. Discovering hidden connections between moral judgements can lead us to extend ideas about one topic to another, as when the valuing of human pleasure leads, by analogical reasoning, to a similar valuing of animal pleasure. Structural theory can even identify wholly new abstract principles, intuitively appealing in themselves and demanding radical changes in our moral outlook. That this approach often starts with everyday moral judgements doesn't mean it's tied forever to them. It can introduce new principles, so long as they imply and explain a set of particular moral claims and use concepts not drastically different from those in the claims.

In pursuing the structural approach we should attend to the richness not only of everyday thought but also of the normative theories philosophers have developed. Philosophy often reduces the main contending views on a subject to a small number: in normative ethics typically just utilitarianism, Kantian deontology, and (a recent addition) virtue ethics. But there are vastly more options than those: non-utilitarian consequentialisms such as Moore's, non-Kantian deontologies such as Ross's, and rival accounts of virtue like the one described above. We should also attend to the full richness of our philosophical tradition. When contemporary philosophers align themselves with historical figures they tend to pick one from a small group: Aristotle, Hobbes, Kant, and maybe Hume or Sidgwick. But again there are vastly more writers worth reading, including especially, I would say, those early $20^{th}$-century figures such as Moore and Ross who developed moral theories starkly opposed to Aristotle's and Kant's.

A recommendation of structural over foundational theorizing is, I recognize, unlikely to have much effect; the latter is too attractive to philosophers. And it would be going too far to say foundational work is never of value. Even when its projects fail, the efforts devoted to them can in various ways be instructive. But surely even those drawn to it need to do structural work first. If they're to give deep justifications of our moral judgements, they must first know exactly what those judgements say. And if there are structural principles that partially explain our judgements, a more foundational theory should adopt them and incorporate their claims. So even those drawn to the other approach should see structural theory as a necessary propadeutic.

It doubtless looks self-serving: a philosopher recommends as the most valuable approach to moral theory the one he himself has pursued. But in some cases the causation runs in the opposite direction: a philosopher first feels, even in the inchoate way I did in my undergraduate seminar, that a certain style of theorizing is most fruitful, and is then drawn to practise it by that feeling. And my recommendation is by no means unique to me. I close by quoting a passage from Friedrich Nietzsche, urging a similar approach while directing appropriate barbs at Kant and his followers:

> One should own up in all strictness to what is still necessary here for a long time, to what alone is justified so far: to collect material, to conceptualise and arrange a vast realm of subtle feelings of value and differences of value which are alive, grow, beget, and perish ... all to prepare a *typology* of morals.

> To be sure, so far one has not been so modest. With a stiff seriousness that inspires laughter, all our philosophers demanded something far more exalted, presumptuous, and solemn from themselves as soon as they approached the study of morality: they wanted to supply a *rational foundation* for morality – and every philosopher so far has believed that he has provided such a foundation. How remote from their clumsy pride was that task which they considered insignificant and left in dust and must – the task of description – although the subtlest fingers and senses can scarcely be subtle enough for it (Nietzsche, *Beyond Good and Evil*, sec. 186).

## Bibliography

Broad, C. D. 'Self and Others,' in D. Cheney, ed., *Broad's Critical Essays in Moral Philosophy* (London: George Allen & Unwin, 1971).

Hurka, Thomas. *Perfectionism* (New York: Oxford University Press, 1993).

———. *Virtue, Vice, and Value* (New York: Oxford University Press, 2001).

———. 'The Common Structure of Virtue and Desert,' *Ethics* 112 (2001).

Hursthouse, Rosalind. *On Virtue Ethics* (Oxford: Clarendon Press, 1999).

Kagan, Shelly. 'Equality and Desert,' in L. P. Pojman and O. McLeod, eds., *What Do We Deserve? A Reader on Justice and Desert* (New York: Oxford University Press, 1999).

Kamm, F. M. *Morality/Mortality, Vol. II: Rights, Duties, and Status* (New York: Oxford University Press, 1996).

Korsgaard, Christine. *The Sources of Normativity* (Cambridge: Cambridge University Press, 1996).

McMahan, Jeff. 'Killing, Letting Die, and Withdrawing Aid,' *Ethics* 103 (1993).

Moore, G. E. *Principia Ethica* (Cambridge: Cambridge University Press, 1903).

Nietzsche, Friedrich. *Beyond Good and Evil*, trans. W. Kaufmann (New York: Vintage, 1966).

Parfit, Derek. *Reasons and Persons* (Oxford: Clarendon Press, 1984).

———. 'Equality or Priority?', The Lindley Lecture (Lawrence, KS: University of Kansas Press, 1995).

Prichard, H. A. *Moral Obligation* (Oxford: Clarendon Press, 1948).

Rawls, John. *A Theory of Justice* (Cambridge, MA: Harvard University Press, 1971).

Ross, W. D. *The Right and the Good* (Oxford: Clarendon Press, 1930).

Scheffler, Samuel. *The Rejection of Consequentialism* (Oxford: Clarendon Press, 1982).

Temkin, Larry S. *Inequality* (New York: Oxford University Press, 1993).

# 7

# Jeff McMahan

## Professor of Philosophy

Rutgers, The State University of New Jersey, USA

---

**Why were you initially drawn to normative ethics?**

How does one explain an interest in ethics? In my case the interest has never been "intellectual" or "academic." I have never been drawn to metaethics. Rather, I have always been aware that there's a lot wrong in the world and I have wanted to do what I could to help put it right. I grew up in the American south during the years of the Vietnam War and the civil rights movement. That gave me a lot to think about. I still have a poster that I took off a telephone pole near where I lived in rural South Carolina that reads "Don't be half a man, join the Klan."

Although I have sought to contribute to the mitigation of some of the world's evils, I have also wanted to be sure that I know what I'm doing. Many of the most horrendous crimes in history have been done or instigated by people who were convinced that they were acting nobly, in accordance with the stern demands of morality. Much of the injustice I saw when I was young was done in the name of religion. (*Plus ça change, plus c'est la même chose.*) I had received routine indoctrination in the Presbyterian Church, which I attended regularly for Sunday school and church service throughout my childhood—albeit under duress, for I was never a believer. It's not that I was a precocious contrarian. I wasn't; I had an ordinary childhood, never read books, and was a very poor student. It's just that religion didn't make sense to me. Later on, when what Swift described on his tombstone as "savage indignation" drove me to political activism, I wanted to make sure that I didn't follow various religious zealots in making things worse through the complacent acceptance of meretricious beliefs.

The tendency to Swiftian indignation is one of many dispositions with which I'm burdened that, especially nowadays, are found unattractive. I am "critical" and "judgmental," cynical and misanthropic. I become choleric when reading the newspaper, and indeed whenever I have to leave the house I'm barraged with little instances of the thoughtlessness, mean-spiritedness, selfishness, obtuseness, irrationality and cruelty. The national elections of 2004 didn't help.

It is these unappealing dispositions of temperament, combined with an aversion to crusading moralism based on delusion, that ultimately led me to philosophy, and to normative ethics.

I am not exempt from my own critical tendencies. When I was in junior high and high school, I spent as much of my time as I could killing birds with a shotgun. Eventually revulsion with myself and what I was doing overcame my unaccountable pleasure in this activity and I sold my shotgun and vowed never to go hunting again. Then I had one of my earliest philosophical thoughts: if I ought not to be killing animals in order to eat them, I ought not to be paying other people to kill them so that I could eat them. I decided to become a vegetarian, which I did when I went off to college and could exercise some control over what I ate. I have remained a vegetarian, though I now realize that there are arguments against eating meat that are much stronger than the one that initially persuaded me.

I went to college with the intention of majoring in studio art. But I immediately fell under the influence of an elderly, avuncular, and wise professor of English literature, and became an English major instead. This was well before the ascendancy of "theory" and cultural criticism in English departments; so I actually spent my time reading literature. The professor whom I venerated was a moralist in the sense that what he mainly looked for in literature was moral wisdom. And what he found he offered to his students.

The department in which I studied English was really a department of world literature. I became enamored with poets and novelists who traded in ideas: Lucretius, Shelly, Dr. Johnson, Voltaire, Tolstoy, Dostoyevsky, and Huxley, among others. Somehow I also discovered Bertrand Russell, who became my first hero. I read his book, *Has Man a Future?* and Linus Pauling's *No More War!*, both of which dealt with the threat of nuclear war. They had been published quite a bit earlier but the weapons had not gone away. These books left me fearful of the destructiveness of irrationality.

As I continued to study literature, I became increasingly impatient with ideas decorated in literary form. I wanted to be as sure as I could be that what I believed was defensible. I wanted to examine the arguments. After completing my BA in English in the US, I therefore took up the study of philosophy at Oxford. I had, however, very little coursework or formal training – only four 8-week undergraduate courses in philosophy before going on to write my doctoral dissertation. My competence in philosophy is, as a consequence, lamentably narrow. But so are my interests in philosophy, so I've never seriously endeavored to remedy my deficiencies. (I recall my eventual dissertation supervisor, Bernard Williams, saying to me once that he didn't think that anyone could do ethics competently without a thorough grounding in logic. I nodded solemnly as if to register agreement, though I had never spent a minute studying logic and didn't even know what modus ponens was – in fact, I still don't, though I know it has something to do with p and q.)

When I was doing this remedial undergraduate work at Oxford, ordinary language philosophy was moribund but still had enough of a pulse to make most of what I was studying seem arid and unimportant. I was on the verge of bailing out of philosophy altogether when I was rescued by discovering the work of Jonathan Glover and Peter Singer. I knew when I read Glover's *Causing Death and Saving Lives* that the issues it addressed – the life-and-death issues, such as abortion, war, and capital punishment – were issues that mattered and were what I wanted to work on. And by and large that is what I have subsequently done.

Shortly after I began my dissertation work (initially under Glover, then under Derek Parfit, and finally, when I transferred from Oxford to Cambridge, under Williams), the Soviet Union invaded Afghanistan, Ronald Reagan became president, and NATO revealed a plan to deploy nuclear-armed cruise missiles near where I was living. My wife and I became activists with the Campaign for Nuclear Disarmament and I wrote a short book on British nuclear weapons policy, arguing the case for unilateral nuclear disarmament. At this point I discovered the political writings of Noam Chomsky and the scales fell from my eyes. (I retain quite a collection of scales, however, and they continue to fall with some regularity.) Chomsky enabled me to see, among many other things, that one of the most useful functions of nuclear weapons was to deter opposition to unjust military interventions in countries that the US was competing with the Soviet Union to control. While I continued to pursue my graduate work in philosophy, I wrote another book on the Reagan administration's foreign policy, which focused both on its nuclear weapons policies and its various interventions in the third world, primarily in Central America.

The two books that I wrote while doing my graduate work were not philosophical but were primarily concerned with politics and policy. The one on Reagan was polemical in character and, unlike my philosophical work, was in many ways enjoyable to write. Both books were, again unlike philosophy, rather easy to write, and neither had any discernable impact on anything. Yet they were concerned with the issues that I thought really mattered. Without discouraging me from writing them, Parfit warned me that to the extent that they took time away from my philosophical work, they would hinder rather than advance my prospects for a career in philosophy, and in retrospect I believe he was right. But I have no regrets, or at any rate not many.

By the middle of the 1980s I had begun to write on the ethics of nuclear deterrence and international intervention in a more philosophical way. And all the while I continued to work on issues in which my interest had been kindled by reading Glover and Parfit, by studying with them, and by reading Singer: issues such as causing people to exist, abortion, infanticide, euthanasia, the moral status of animals, killing in self-defense, conventional war, capital punishment, and so on.

I no longer go to rallies or on demonstrations and may never again write political tracts of the sort I wrote as a graduate student. I have ceased to be an activist, mainly because there has to be a division of labor and I'm certain that I can make a more substantial contribution by analyzing, criticizing, and formulating arguments than by waving placards and circulating petitions. Most people who write and debate about the most important moral and political issues are not, to put it as tactfully as I can, very skillful or careful in their reasoning. Having trained in philosophy, however inadequately, in order to be able to reason and argue with clarity and rigor, I am trying to get the arguments right with respect to some of these important issues. This has taken me deep into the territory of normative ethics. I have no illusion that my work will ever be as widely read or influential as that of Glover, Singer, and others who have inspired me to follow the path I have taken. But I continue to be motivated by the same concerns that initially led me into this curious vocation.

**What example(s) from your work (or the work of others) illustrates the role that normative ethics ought to play in moral philosophy?**

My hope is that all of my work does this, at least to some extent. Of course, normative ethics doesn't have only one role in moral philosophy, but one of its undoubted roles is to enable us to determine how to act permissibly and to live well and wisely. All of my work is aimed at this broad goal. How well it succeeds is for others to judge.

Normative ethics is often thought to lie between metaethics and practical ethics. It is thought to be concerned with how we should reason in a general way about issues in ethics: for example, what is the correct moral theory, and how do we determine what the best moral theory is? Those who think of normative ethics this way often refer to what I call practical ethics as "applied ethics," since they assume that once we have the right moral theory, we can resolve practical problems by simply applying the theory to the problems. According to this approach, theoretical normative ethics has priority over practical normative ethics: we have to get the theory right first in order then to resolve the practical problems. A majority of the most eminent thinkers in normative ethics work this way. They devote most of their efforts to working out and defending a moral theory – some version of contractualism, Kantianism, or consequentialism, or a theory of rights or virtues – and then, mainly in order to clarify the elements of the theory and to illustrate its power, explore some of its implications for a limited range of practical moral or political problems. Prestige in moral philosophy tends to be awarded to those who work this way rather than to those who work piecemeal on particular problems and issues.

I am, however, somewhat skeptical of this approach. I worry that it may be premature to think that we are in a position to develop a plausible moral theory. It is disconcerting, for example, to find people emerging from graduate school thinking that they already know the right moral theory and setting forth to refine it in original ways and to defend it against all comers. I worry that in normative ethics we don't really understand our basic material well enough yet. I therefore think of what I do as rather humble spade work in ethics that may better enable our successors to develop a more adequate moral theory than any on offer. I share the ambitions of systematic moral theory but am less optimistic than many about how far along we are in realizing those ambitions.

Of the work that has been done over the past 100 years, the book that best exemplifies the virtues that I admire in normative ethics is Derek Parfit's *Reasons and Persons*. It neither explores, develops, nor applies a moral theory, but instead addresses a set of apparently unrelated issues and in the process produces a range of startling insights, shows that we have been mistaken about the foundations of some of our most deeply held beliefs, reveals other deep problems of which we had hitherto been unaware, opens up new areas of inquiry, and in the end demonstrates that what seemed to be disparate problems converge in supporting the view that impersonal reasons and values, which a great many moral theorists have denied altogether, are implicit in and presupposed by much of our common moral thought.

### How do studies within scientific disciplines contribute to the development of normative ethics?

I would like to be able to say something outrageous here, such as that ethics is an autonomous domain and that science has nothing to contribute to it whatsoever. But that's false and my answer has to be rather pedestrian. Science contributes to ethics in many ways but, perhaps paradoxically, the most important way may be that by presenting us with new problems and choices, it forces us to think more deeply about ethics than we had before. Our enhanced understanding of embryology and the development of surgical techniques that have enabled us to perform abortions with less risk to the pregnant woman than is involved in childbirth have compelled us to rethink many traditional beliefs about procreation and the value of human life. By showing that human nature is potentially radically malleable advances in genetics have challenged traditional normative methodologies that seek to ground ethics in some specific conception of human nature. Advances in agriculture, transportation, and mass communication have forced us to confront hitherto unappreciated questions about how much we may be required to sacrifice for those less fortunate than ourselves. Other advances in agriculture and nutritional science have made it possible for many of us to live fully healthy lives without eating meat and have therefore prompted us to reconsider traditional views about the moral status of animals. The invention of nuclear weapons has made it imperative to try to understand the importance of the continued existence of the human species. Further examples could be multiplied almost endlessly.

## What do you consider the most neglected topics and/or contributions in normative ethics?

This is an easy question for me because one of the most unaccountably neglected topics in the history of ethics is the topic on which most of my own work now concentrates: the ethics of war. Compared to novelists, poets, historians, legal theorists, psychologists, economists, and even artists, philosophers have had almost nothing interesting to say about war. Yet this is hardly an insignificant topic, or one that fails to raise important and interesting ethical questions.

Historically, the literature on the ethics of war has been dominated by theological and juridical writers, such as Augustine, Aquinas, Vitoria, Suarez, Gentili, Grotius, Pufendorf, and Vattel. There is a little on war in Locke, Rousseau, and Sidgwick, and a bit more in Hobbes and Kant. Yet none of these philosophers ever engaged in sustained reflection on what is morally most conspicuous about war – namely, that it involves the mass killing of people of whom one has no personal knowledge at all. This neglect is really quite extraordinary, and to me very baffling. How can it be that from the Greeks on, philosophers have found that the mass slaughter of one group of people by another demands neither criticism nor justification? How can they have all been so complacent about this most barbaric of activities?

During and after the Second World War, in which at least 70 million people died, entire cities were intentionally bombed to rubble and their populations murdered, and attempts made at the extermination of entire human groups, what did moral philosophers say in their professional capacity? With one honorable exception – Elizabeth Anscombe, who argued on the basis of Catholic just war principles against Oxford's awarding an honorary degree to US president Truman, on the ground that he had ordered the terrorist mass murder of Japanese civilians – moral philosophers were entirely mute about these matters. They were agitated instead about such questions as whether moral propositions had a cognitive component or were merely expressions of approval or disapproval. I think that as moral theorists we should look back on this episode with shame at the dereliction of our forebears – though I stress that I refer only to professional dereliction, as some moral philosophers did fight and suffer in that war.

In the 1970s, the Vietnam War prompted a flurry of interest in the ethics of war, primarily, and for obvious reasons, among American philosophers. One significant work emerged: Michael Walzer's

*Just and Unjust Wars*, in 1977. During the 1980s a modest literature developed on ethical issues raised by the practice of nuclear deterrence, though for the most part these issues failed to engage the attention of major writers in ethics. The Gulf War produced a few philosophical papers and in the late 1990's the various Balkan wars and, perhaps, a lingering sense of shame over the genocide in Rwanda provoked a revival of interest in the theory of the just war, which has continued and grown more vigorous in response to the terrorists attacks of 9/11, the Bush administration's "war on terror," and the disastrous war in Iraq. Much of the recent literature focuses, again for obvious reasons, on the ethics of humanitarian intervention and preventive war, and it has benefited from a parallel stirring of interest among legal and political theorists in the international law of war. My own sense is that some of the work that is currently being done on the ethics of war is unprecedented in its sophistication, rigor, and attention to detail. It is rivaled only by the best passages in the work of Grotius and a few others, and I think it is not unreasonable to hope that contemporary theorists of normative ethics will soon produce a body of philosophical writing on war that will be significantly more illuminating than anything that has been written in the past.

## What are the most important problems in normative ethics and what are the prospects for progress?

There are so many problems that are all so important that I really don't know how to rank them. Among the issues that I personally find most important are problems concerning causing people to exist, such as Parfit's Non-Identity Problem, and the problem of determining whether the distinction between doing and allowing, and the distinction between intended effects and foreseen but unintended effects, have moral significance, and if so, exactly what kind of significance. These problems are of tremendous theoretical and practical significance.

I am highly optimistic about the prospects for progress in normative ethics. It is evident to me that great progress has already been made since I entered the field in the early 1980s. Unlike many other disciplines in the humanities and social sciences, which in recent years were seduced by bad French philosophy into a lot of silly "post-modern" theorizing that has exposed them to derision and reduced them to irrelevance, analytic philosophy is flourishing. Part of the reason why analytic philosophy generally is in such a healthy state is that, as Jerry Fodor observed in a recent book review, philosophers no longer tend to have philosophies. We no longer devote our lives to developing comprehensive philosophical or ethical systems. We are individually narrower and more specialized, which enables us to focus more carefully and minutely on the problems we study, and as a consequence to produce work that is more rigorous and detailed. The result is that philosophy has become more of a collective endeavour than it was in the past, in the sense that different people are focusing selectively on problems that are elements or aspects of larger problems. When the results of the individual efforts are combined, we may achieve a collective product that exceeds in depth, intricacy, and sophistication what any individual could have produced by working on the larger problem in isolation.

# 8

# Jan Narveson

## Professor Emeritus
University of Waterloo, Canada

---

**Why were you initially drawn to normative ethics?**

At the personal level, I think it likely that the main influence here was, first, a family environment in which my parents were active in school (my father was an educator—a superintendant of schools in a small town in my early days, later teaching at a Lutheran liberal arts college) and church (both parents were very active, in all sorts of nonclerical capacities, to the point that after retirement, they spent two years as lay missionaries in New Guinea). There was pressure on the five children in my family to go seriously into religious work, and that pressure had to be dealt with. But especially after moving to a larger town with two quite different colleges in it (one secular, one Lutheran), a more diverse population and new friends, I began reading seriously in the broad literatures of our era and before, and no doubt in part because of that, to bring religious beliefs into close scrutiny. I rejected quite thoroughly the specifically religious part of my upbringing, but I am sure that its general moral influence was considerable. My university career, which has from graduate school onward been almost entirely concerned with moral and political theory and practice, is perhaps a consequence of that upbringing more than any other identifiable factor that I can think of.

Normative ethics as a *subject* is interesting because of its conceptual status; more precisely, it is interesting because it is puzzling just what its conceptual status is. When I turned to serious graduate-level work in ethics, I became aware that the half of the 20th century that had by then elapsed was spent puzzling about that very issue; interestingly enough, we are still puzzling about it. Meanwhile, of course, at the *practical* level, it seems obvious enough that there were and are various things wrong in the world, making it worth while to try to identify more precisely what *is* wrong in it—even if there is so little that any one of us can do. But perhaps getting the theory right may help, at least a bit.

## What example(s) from your work (or the work of others) illustrates the role that normative ethics ought to play in moral philosophy?

This question is rather unclear to me. Since I am a moral philosopher, the "role" that normative ethics plays, as a subject, is simply that it is part of the subject. I have published one book concerned entirely with various general issues in normative ethics, and my others all discuss it in varying ways and amounts.

One way to construe this question bears very centrally on the nature of the subject. Almost everyone who becomes interested in moral philosophy does so against a background of moral beliefs of various kinds. Many of the great figures in moral philosophy – Aristotle, Hume, and in our own day John Rawls come to mind – have been of the view that these background beliefs play a major role in the development of the theory of the subject. Some say that philosophy can't really do more than systematize and harmonize these beliefs, rather than either rejecting them or even inquiring seriously into their foundations. I am strongly of the opposite view. All normative beliefs, at all levels of generality, can be "questioned" in the sense of trying to find genuine reasons for them, and if those are lacking, that has serious resonances in normative practice as well.

An interesting area in point is the issue of abortion, which is still of major public concern. Most philosophers, including myself, have seen reason to converge on a very substantially "liberal" view about this subject, holding that the status of the yet-to-be-born is such as to give potential parents the right to abort if they so choose. This is, though, perhaps more the influence of theory on practice than the other way around. Even so, a major incentive to this liberal persuasion is the experiences of women under regimes of heavy restrictions on abortion, for example, as well as more generally the increasing sense, over the past half century especially, that women had not been getting a fair shake in the social system of previous times, and that providing them with real choice on abortion was a considerable contribution to giving them their rightful place in the social order.

As regards the theoretical issue about the role of pre-philosophical moral beliefs, however, I side with the classic view: that we philoso-

phers need to pursue moral philosophy at all levels including the
most fundamental ones, and to draw on our conceptual inquiries
for results that may strongly affect practice in many areas. In that
sense, I am a "top-downer," as one might put it. And yet, as always, one's view of the fundamentals of a field derive broadly from
particulars—from our sense of local issues at many points. It can
hardly be otherwise. But it is easy to misconstrue that influence.

### How do studies within scientific disciplines contribute to the development of normative ethics?

The contribution of science to ethics is undoubtedly huge. That
is especially because science – at least if we construe that term
broadly so as to include correctible commonsense information about
the world – provides the "minor premises" of ethical arguments.
If we hold that such-and-such ought to be done in cases of type
C, then science can help to tell us whether we do actually have
before us such a case. All normative issues have this aspect. (I
reject the idea that it is impossible to disassemble the "thick concepts" of ordinary ethics, as so many of today's writers seem to
think.) It is even possible that some developments in, say, neuroscience could tell us what's going on in the brain when people
make ethical choices, etc. Can that make a difference to theory?
It is rarely safe to generalize about the longer-run prospects of
science, but it would be foolish to doubt that the further development of many sciences will have major effects on many important
moral questions, including even very high-level and basic ones.

### What do you consider the most neglected topics and/or contributions in normative ethics?

One could suggest that some issues have not been neglected nearly
enough of late! But in general, the amount of work being done
on just about every normative issue one can imagine is immense
these days. It was not always so. Going further in response to this
question runs an obvious danger of sectarianism—segueing from
"not enough is being done" to "the wrong things are being said."
Insofar as we can quantify without taking sides, I would be hard
put to try to identify any *subject* that has been really "neglected."
There is, of course, always room for more good thinking about
anything, but that's not to say that people aren't thinking about
these things at all. They are, and often very effectively.

## What are the most important problems in normative ethics and what are the prospects for progress?

Game-theoretic studies of human interactions providing scope for cooperation, are the most fundamental inquiries in ethics. But it needs to be construed very widely so as to include the study of people acting under the influence of various normative ideas. At present, there is a special need to consider the question of what to do about possible threats and dangers of a very, very long-term nature. An important current example is "global warming." Besides all the scientific uncertainty about the causes and extent of this phenomenon, we need, very seriously, to address the question of what to do NOW about possible problems that develop very, very slowly over very long periods of time, especially when the knowledge-environment is changing so rapidly. The likelihood that we will be in a position to bring new techniques to bear, say, fifty years from now, is enormous. We may be wasting resources if we try to do something about them now. A favourite example of mine is the Kyoto accords, whose policy recommendations seem to me to be complete nonsense in the present situation: Scientists have found that the impact of adhering to Kyoto over fifty years would be on the order of 0.1 C. of reduction in global temperatures— a completely insignificant, indeed undetectable reduction, got at huge cost to billions of people. That is not progress—it's grand larceny on the part of governments.

Far more important than this is the problem of how to deal with people engaged in what looks like large-scale murder in the interests of religious fanaticism. That is a long-standing problem, but more serious today than perhaps at any time in the recent past, at least. In centuries gone by, however, one can point to periods such as the Crusades, and before that the era of Muslim conquest, when religiously-inspired warfare was absolutely the norm. But the greatly increasing amount of interaction among diverse peoples, combined with developments in military technology, make this into arguably the most critical issue of the day. What's interesting about this particular matter is that philosophical inquiry is so centrally relevant. Those who would make war on others because of religious conviction have a position that is impossible to defend coherently, and the need for that to be generally perceived seems to me to be very pressing today. Of course, the need to understand also the sources, psychological and sociological, of that kind of fanaticism is very great as well. But it strikes me that pure philosophy is very, very important to this area.

Finally, I would suggest that the ongoing interest in specifically political philosophy is also of greater importance than ever today. There has been much writing about democratic theory, but almost all of it suggests the extreme limitations of pure democracy as a political form. Democracy without serious constitutional restraints that are seriously respected by governments and populations is another of the great enemies of our time. Pressure to democratize is great, but succumbing to it without adequate thought about its constitutional constraints is likely to be fatal.

# 9

# Onora O'Neill

Professor of Philosophy

Cambridge University, UK

## Why were you initially drawn to normative ethics?

I am unsure whether I have ever been drawn to normative ethics. This may seem surprising, in that I have often commented on normative issues, including ethical ones, in my writing. In responding to the questions set by Thomas Petersen and Jesper Ryberg, I have been led to think more closely about ways in which I have and have not engaged in the sort of work usually spoken of as normative ethics. My interests have always been in underlying issues about reason and action, and in the relation between them. I am more fascinated by normativity and its presuppositions than I am interested in ethical or other substantive norms. My reflections are arranged as a narrative, but the underlying points are systematic.

As a graduate student at Harvard in the late 60's I found myself wondering whether and how reason could be practical. I took a seminar with the late Robert Nozick, in which we read *Games and Decisions: Introduction and Critical Survey* by R.D. Luce and H. Raiffa. It was my first solid encounter with game theoretic and formal approaches to reasoning about action, and initially I was rather taken by their neatness—I even started working on ways in which interpersonal utility comparisons might be made, despite the apparent lack of a metric. Bob Nozick was kind enough to say that he liked what I was writing and to encourage me to publish it—but he had hardly done so when I had a moment of revulsion. I think that I concluded simply that if a conception of practical reason needed quite so many implausible assumptions, something was going deeply wrong. Although I did not work it out fully until later (in *Faces of Hunger*, 1986), I came to think that consequentialist practical reasoning was doomed either to pleasant vagueness or to reliance on illusory metrics, so it was unsatisfactory.

Like many before me I went back to Kant. I aimed to find out what his conception of practical reason had to offer. Initially I approached Kant through the lens of mid twentieth century writers – for example, M.G. Singer and R.M. Hare – who had suggested various ways in which quasi-Kantian notions such as universal generalization or universal prescriptivism might be used in practical reasoning. As I came to see how heavily they too relied on quasi-consequentialist frameworks, I began exploring Kant's work more thoroughly. Harvard was an excellent place for these explorations. I took Charles Parsons' course on the *Critique of Pure Reason* (which left me with an abiding interest in the *Doctrine of Method*); I struggled with *Religion within the Limits of Reason* when I was a teaching assistant for Stanley Cavell; I attended John Rawls's careful and illuminating lectures on Kant's ethics, and it was under him that I wrote my Ph D. I set myself to show that the Categorical Imperative could guide action, and so perhaps provide a basis for at least some forms of normative claims that were relevant for all.

At this early stage of my work I did not try to explain *why* we should think of the Categorical Imperative as a principle – let alone *the principle* – of practical reason. I was preoccupied with trying to show that there was at least one interpretation of the Categorical Imperative that could ground at least some normative claims about human duties. I did not try to extend this work into substantive normative claims of any specificity. The work I was trying to do led me to think a lot about the idea that ethically acceptable principles of action must have a certain form, about Kant's notion of a maxim, and more broadly about the conception of action needed for practical reasoning. At this time I took a fairly traditional view of maxims as internal states of individuals at and through certain times. I argued that it was essential to focus on the propositional structure of maxims, in order to reach a coherent interpretation of Kant's claim that obligation is a matter of refraining from action on principles that cannot be principles for all. It was only in the 1980s that I started to question this individualistic view of maxims.

**What example(s) from your work (or the work of others) illustrates the role that normative ethics ought to play in moral philosophy?**

Normative ethics did not play much part in my own work until the 1970's. In this I was not at all unusual. Many philosophers then assumed that while examples of right and wrong action could be cited to *illustrate* philosophical arguments about duty or right, developing an account of substantive, including normative, ethics was not a philosophical task. Metaethics could be approached philosophically; substantive ethics demanded much that philosophers were not well equipped to provide.

However, the political upheavals of the 1970s propelled many of us to take a more engaged view of the ways in which philosophical writing could bear on normative issues, and like many others I began to teach and write in a different way, that often discussed normative questions. However, for reasons I'll try to explain, I did not aim to write in normative ethics, and (for reasons I shall explain) have remained quite sceptical about the prospect of 'applied' ethics.

The normative questions that I started to refer to in the 70's were generally political: I tried to develop arguments that would be relevant to action that bore on poverty, to conceptions of equality and equal opportunity (still in my view a morass of unclarity) and to claims about liberty. While teaching in NYC in the mid 70's I also encountered a wide range of normative work, and was quite involved with the *Society for Philosophy and Public Affairs*. Bill Ruddick and I organised a series of seminars for the Society on the ethics of parenthood, and then edited a book under the title *Having Children: Legal and Philosophical Reflections on Parenthood* (1979). Even at that time we noticed with some regret how much questions about rights were coming to dominate work in normative ethics. It was also at this time that I first became aware of the rapid emergence of medical ethics, which has played so large a part within normative ethics.

However, normative ethics continued to play a relatively small part in my own writing. I continued to write more on Kant and on political philosophy than on ethics, and what I wrote on ethics in the 70's and 80's was mainly on some of the presuppositions of normative ethics – in particular on principles, abstraction, obligations, rights and virtues – rather than on substantive normative ethics. I developed a series of arguments for thinking that obligations are more fundamental than rights, and that arguments for rights need to be based on claims about obligations.

In this I have often felt rather isolated: I suspect that rights, and specifically human rights, are so often seen as fundamental to normative ethics mainly because claiming rights is much more alluring than living up to obligations. To this day I have not met convincing arguments for treating rights as foundational to ethics. I have never been hostile to rights, or to human rights, but I think that they deserve intellectually more robust foundations than are provided by those who assume that we do not take rights seriously unless we view them as normatively fundamental. I took increasingly to illustrating underlying arguments about the relation of obligations to rights, and about the importance of agency, with extended discussions of specific obligations and rights. In this way I came to publish on children's rights, on gender and rights, on justice and charity, and on justice and virtue, and so (I think) can pass as having done a good bit of substantive normative ethics. However, I chose these topics as illustrative of certain underlying arguments and lines of thought, and the arguments that interested me remained those of moral philosophy. The centre of gravity of my work remained in accounts of reason and action, rather than in normative ethics.

### How do studies within scientific disciplines contribute to the development of normative ethics?

I would contrast the approach that I have taken with approaches to normative ethics that are based on deep knowledge and understanding of specific empirical and scientific literatures. I greatly admire those who master other areas of inquiry so fully that they can incorporate a rich range of empirical detail into their discussion of normative issues. I have never aimed to do so, although I am sufficiently inquisitive to read quite widely beyond philosophy, and to my pleasure find that I can talk fruitfully with practitioners in many areas. However, I think I have learned a good deal more about the full range of normative issues by taking on practical commitments in a fairly wide range of institutions, by involvement in a range of bioethics committees and by lecturing to wider audiences, than I have by studying other disciplines or by writing in normative ethics.

In the mid eighties I returned to fundamental questions about practical reason. Since then I have tried to show not only that certain principles – among them the Categorical Imperative – can be action guiding, but that there are grounds for thinking of them as principles of reason. I have been increasingly concerned not only with showing what it takes for principles to be practical, i.e. action guiding, but with articulating what it takes for them to be reasoned. This focus has led me to think a lot about the assumptions made by various contemporary political philosophers whose work is seen a contractualist, or more broadly as constructivist, and about various conceptions of 'public' reason. In working on these topics I have moved away from my early individualistic, indeed mentalistic, conception of principles of action. I now see practical reasoning as bearing on principles which agents can seek to enact, but which they may or may not consciously formulate. Practical principles, as I see that matter, are standards to which we can seek to match action, rather than states of agents. The fundamental issue for all forms of practical reason is that its direction of fit differs from that of theoretical reasoning: practical reasoning aims to shape the world (in some small part) to fit certain principles; theoretical reasoning aims to fit principles to the way (some aspect of) the world already is. This is why I do not think it is helpful to talk about 'applied ethics': we *apply* concepts and principles to aspects of the world in theoretical reasoning, often with the aim of understanding or explaining. But in practical reasoning, including ethical reasoning, we aim to *enact* or *instantiate* principles, and so to change the world in at least some small way.

This renewed preoccupation with the nature of practical principles led me to think more about their scope. Evidently some practical principles are of narrow scope: they are relevant only to closely specified types of agent. But others may be thought of having a wider scope, perhaps one that extends to all or many reasoning agents. Yet establishing the scope of ethical norms remains uphill work and I found myself engaged in a range of discussions about the scope and limits of justice in the 80s and 90s. These discussions have manly taken place among political philosophers, and in particular among those who are interested in discussions of justice that goes beyond the borders of states or communities. Those who think that the scope of justice can be limited to others with whom we share a community, state or ideology (not only communitarians, but many liberals) have generally been opposed by those who take more cosmopolitan positions. I have generally found myself agreeing with the underlying views of cosmopolitans, but also arguing against the idea that justice requires a world of open borders or a world state: both, as I see it, would risk concentrating too much power and incompetence. Even if the scope of justice is wider than any existing state, justice may require the preservation rather than the abolition of boundaries.

Given that the requirements and scope of practical reasoning remain fundamental to my work, I am quite surprised to realise that I have in fact written a bit in normative ethics, and in particular on bioethics. For many years I had hesitated to write on normative issues except in order to illustrate what I took to be underlying arguments. Given my views on practical reasoning, I found it hard to see how I could write anything philosophically satisfactory in normative ethics, or in particular in bioethics. However, I came to think that there were ways of doing so that did not beg the underlying questions about action and reason that had come to preoccupy me. Since 2000 I have surprised myself by writing two books that can pass as normative ethics: *Autonomy and Trust in Bioethics* (2002), and jointly with Neil Manson *Rethinking Informed Consent in Bioethics* (2007).

## What do you consider the most neglected topics and/or contributions in normative ethics?

A few years ago I concluded that the most neglected topic in normative ethics was *trust*—and the most overworked topic a certain range of individualistic views of *autonomy*. I explored both themes in my 2002 Reith Lectures, published as *A Question of Trust*, as well as in the two works on bioethics just mentioned. A standard view is that trust is merely an attitude that may be directed not only at particular individuals or institutions, but at *types* of office-holder or institution. Supposedly empirical claims about trust – often about declining trust, or failing trust, or even about a crisis of trust – are often based on no more than studies of attitudes towards those holding certain offices, such as those provided by opinion polls. I came to the view that, on the contrary, trust is a matter of judgment and is typically based on an assessment of some – although not of conclusive – evidence. The epistemic basis of trust seems to me an important and still neglected topic, and this neglect has deeply damaging practical consequences.

Once trust is misconstrued as a mere attitude, to be discovered by asking people to rate their attitudes or feelings of trust towards types of office holder (police, doctors, nurses, politicians, journalists, etc.), it seems that that there is no practical remedy for loss of trust. Attitudes may or may not reflect available evidence, and may persist in the face of contrary evidence. It is often

easy to maintain an attitude of mistrust despite evidence of trust-
worthiness, and sometimes possible to maintain an attitude of
trust despite evidence of untrustworthiness. Those who see trust
merely as an attitude have few practical proposals for dealing with
the crisis of trust that they often diagnose and deplore. Some of
them end up recommending the abandonment of trust, and even
recommend a general attitude of mistrust and suspicion.

Yet refusing trust because it may sometimes be misplaced is
not an adequate response to the variety of practical situations we
face. We have in the end to work out where to place and where
to refuse it, and we are likely to do so better if we attend to
the evidence. Equally those who hope that others will trust them
need to attend the types of evidence that can support others in
making judgements about where to place their trust. Trust cannot
be extracted from others, but intelligently designed institutions
and practices can make it easier of people to judge where to place
their trust.

## What are the most important problems in normative ethics and what are the prospects for progress?

I suspect that there can be no timeless list of 'the problems of
normative ethics'. The problems that people face, the sorts of
temptations to act badly or wrongly against which they may (or
may not!) struggle, and the opportunities and capacities they may
have to act well or do right, are endlessly varied.

Given that practical problems – ethical and other – vary for
different people and in different times and situations, it may not
make sense to speak of progress quite generally. We can talk about
progress in mathematics, or in understanding a given disease—
but where people have quite often to deal with new or different
problems, it may not make much sense to ask whether normative
ethics *as a whole* has made progress. At most it may be reasonable
to ask whether there has been or could be progress in dealing with
resolving, continuing or recurrent practical problems.

# 10

# Ingmar Persson

## Professor of Practical Philosophy

Gothenburg University, Sweden

---

**Why were you initially drawn to normative ethics?**

When I was 13 or 14, and interested in history, philosophers came across me as odd characters who wandered the streets of Athens. I was immediately enthralled by the fact that they took their doctrines so seriously that these shaped their whole way of living, even when this led to violent conflicts with the conventions of their society. This has ever since been my model of philosophy: philosophy as a way of life. Philosophy offers a unique possibility of combining theory and practice, by seeking an understanding of the most general aspects of reality — time, causation, matter, mind, knowledge, action, will, value, etc. — and basing guidelines of conduct upon this understanding.

Normative ethics in those ancient days was more broadly conceived than it is today. Nowadays, the topic of normative ethics is how we should behave towards beings *other* than ourselves, beings for whom we can make things better or worse. It does not deal with how we should lead *our own* lives. But in antiquity this interest in how to lead one's own life successfully, in view of the fact that doing so is to such a great extent at the mercy of factors beyond one's control, was paramount. In particular, there was a preoccupation with the question of how one should live in the face of death. This is easily understandable, since in those days death was an ever-present threat to a greater degree than in affluent societies today, with modern technology and medicine. A well-known example of this ancient concern is Epicurus's attempt to show that there is no good reason to fear death because, when we are dead, we are not conscious of anything and, so, cannot be conscious of anything that is bad for us, such as pain. As Epicurus took fear of death to be the greatest obstacle to a good life, he thought its elimination could pave the road to such a life.

This conception of normative ethics, according to which was included answering the question of how to live one's own life, held sway for a long time in the history of philosophy. For instance, this seems to be the conception that Spinoza had in mind when he wrote his *Ethics* (1675). But such a conception is not that with which contemporary analytical philosophers work. To understand what they regard as their object of study, I think we had better take common sense morality as our point of departure. According to this morality, we are irrational or stupid, rather than immoral, if we harm our own interests by wasting our talents or by undermining our health. We are seen as being morally permitted to do as we please in the sphere of so-called "prudence", in which only our own interests are affected, so long as we satisfy the conditions of being autonomous, conditions such as being free (or not coerced), rational and well-informed. The broad conception of normative ethics whose object of study comprises how we should handle our own interests no less than the interests of others is currently so unfashionable that there is no adequate term for it in English. In my book *The Retreat of Reason — A Dilemma in the Philosophy of Life*, I used "philosophy of life" to designate it, without being too happy with this choice of terminology.

Although the question of how one should lead one's life has received scant attention by contemporary philosophers (with the exception of existentialists), there is an intricate interplay between it and morality. For instance, as modern medicine develops more and more effective methods of prolonging human life, it becomes imperative to reflect carefully on the question of what makes our lives worth living and abstain from whatever is not essential to accomplish this. If out of an irrational fear of death, we who are inhabitants of affluent countries cling at great expense to lives that are of little or no use to us, we exacerbate an unfair and unsustainable global situation. Generally speaking, the greatest current moral problem is probably that wealthy nations persist in a style of life which consumes so much of the planet's resources that this style could not possibly be shared by the billions in the developing countries.

It is my view that "morality" in the narrower sense, which excludes the domain of prudence, marks a shallow distinction. For I have argued (*The Retreat of Reason*, part IV), following Derek Parfit (*Reasons and Persons*, part III), that the distinction between oneself and others is rationally and, thus, morally irrelevant. It is not hard to understand how this view could be exploited to buttress a proposal to revise morality so that it comes to demand a great deal more from us than is commonly assumed. On the other hand, it should be admitted that there is something paradoxical about expecting people to be enlisted in the moral service of helping others to good lives to such an extent that they themselves are prevented from what they see as leading good lives. I shall return to this dilemma towards the end of this piece.

## What example(s) from your work (or the work of others) illustrates the role that normative ethics ought to play in moral philosophy?

I think that the philosophical discipline of normative ethics should start by trying to lay bare the content of common sense morality. The nature of morality is contested, and the most uncontroversial way of ensuring that we mean roughly the same by "morality" seems to be by taking as our starting-point common sense morality, the morality with which we grew up. Common sense morality appears to contain the following elements.

First, we assume ourselves to have *rights* to things, like our lives, limbs and property. The most general of these rights are *negative*, i.e., they are rights against other persons, or beings capable of recognizing rights, that they do *not* interfere with our use of our lives, limbs and property. They are not rights to receive positive aid from others to help us maintain our lives, limbs and property. The ground of these rights is probably what it has been claimed to be by philosophers in the tradition of John Locke (*Two Treatises of Government*, book II), that is, roughly facts to the effect that we are the first to "appropriate" or "occupy" the things to which we have rights. These *general* rights, which all people have, stand in contrast to the *special* rights which we bestow upon certain individuals by our acts towards them, e.g., the promises we give them. Special rights may well be positive.

Second, common sense morality features some *deontological principles*. There is *the act-omission doctrine*, AOD, according to which it is harder to morally justify the doing of certain harmful actions, such as killing and stealing, than letting them happen. There is also *the doctrine of the double effect*, DDE, according to which it is harder to morally justify the doing of something harmful as an *intended* end or as a means to some greater good than the doing of this as a merely *foreseen* effect of the good. I believe that AOD involves the theory of rights just mentioned (thus, the elements I am distinguishing overlap). This is the explanation of why we are under obligations to refrain from certain harmful acts – they violate rights – but are under no obligations to do good. However, AOD does not boil down to this rights theory, for then our duty to prevent others from violating rights could be as strong as our duty not to violate rights ourselves. As regards DDE, this is best seen as presupposing AOD, as introducing a condition (that the harm is merely foreseen) which allegedly makes doing harm easier to justify morally — in fact, so easy to justify that it can be permissible to *do* a smaller harm rather than to *let* a greater harm occur, e.g., to divert an existing threat so that it kills a fewer number. (See my "The Act-Omission Doctrine and Negative Rights.")

Third, common sense morality includes considerations of *justice or fairness*. These are intimately related to the theory of rights (another overlap), for pre-reflectively we think it just that people retain that to which they have rights, and unjust to deprive them of this. The concept of *desert* also has a central place in our everyday conception of justice: other things being equal, it is seen as just that individuals get what they deserve. But I also think that the concept of *equality* plays an important part in the sense that there is a formal principle of justice to the effect that it is just that individuals are equally well off, unless some deserve to be better off than others or have rights to things which make them better off than others. Consequently, the greater the space for deserts and rights, the lesser the space for equality.

Fourth, common sense morality comprises *reasons of beneficence* to the effect that there is reason to benefit individuals, by helping them to maintain things to which they have rights. According to common sense morality, these reasons are weaker than reasons of rights, but they may on occasion outweigh them: if I can save someone else's life by giving away a few drops of my blood, I ought to do so. But I am not morally required to sacrifice, say, one of my arms to rescue someone else from death, although an arm is of less value to me than life is to the other person. This sacrifice is rather supererogatory.

The space in which these moral reasons do not settle what we morally ought to do could be called the province of *autonomy*. We are there morally permitted to do as we please, provided that our choices are autonomous, i.e., free, rational and informed. The province of autonomy includes a wide range of actions with respect to things to which we have rights. For instance, I am permitted not to sacrifice my arm to save your life, as well as permitted to sacrifice it—indeed, I am even morally permitted to sacrifice my life to save your arm, though your arm presumably is of less value to you than life is to me, provided that I fulfil the conditions of being autonomous.

*Intuitionist* moral philosophers believe that common sense morality is more or less correct as it stands. A prominent exponent of this type of approach is Frances Kamm (see her *Morality, Mortality,* vol. I & II). For these philosophers the task of normative ethics is virtually completed when common sense morality has been detailed. Other philosophers think that common sense morality may contain elements that are confused or unjustifiable and, consequently, that it is in need of considerable revision. I am myself a rather radical *revisionist* in this sense. I believe that general rights, deserts and the two deontological principles all have to be rejected. True morality is the morality we obtain if we cleanse common sense morality of its defective components, and I believe that it will diverge radically from common sense morality. Let me exemplify.

When we reject rights, reasons of beneficence will have to be purged of any reference to them. They will simply be to the effect that we have reason to do what benefits beings, or makes things better for them. But it is then arguable that we could make things better for beings by causing them to exist, whereas this cannot be anything that helps them maintain things to which they have rights, since the non-existent cannot have any rights. Thus, one possible revision of common sense morality is an extension of the range of morality to include possible beings that could be caused to exist. This extension raises problems of a sort discussed, for instance, by Parfit in *Reasons and Persons,* part IV, and by myself, e.g., in the following articles: "Genetic Therapy, Identity and Person-Regarding Reasons", "Person-Affecting Principles and Beyond" and "Rights and the Asymmetry Between Causing Bad and Good Lives".

In *The Retreat of Reason,* chapter 34, and in my paper "A Defence of Extreme Egalitarianism" I argue that there cannot be anything, like deserts and rights, which makes it just that some of us are better off than others, unless there is something for which we are *ultimately responsible*. We are *directly* responsible for our actions in virtue of our intentions and beliefs that make them intentional. We are *ultimately* responsible for our actions if and only if we are responsible for all the conditions in virtue of which we are directly responsible for these actions. Temporally finite agents like ourselves cannot however be ultimately responsible for anything: if the world is deterministic, the states in virtue of which we have direct responsibility will be determined by causes which preceded our existence and, thus, will be beyond our responsibility, whereas if the world is indeterministic, these states will be (partly) undetermined and random, and to this extent beyond our responsibility. In neither case is there a foothold for anything, such as deserts and rights, which is designed to make it just that some of us are better off than others. Since this is so, the formal principle of justice mentioned above implies that a situation is just if and only if we are all equally well off in it.

In "A Basis for (Interspecies) Equality" I contended that this egalitarianism covers not only human beings, but all sentient animals, irrespective of biological species. This is so because the limitations in respect of capacity for well-being that are a consequence of species membership cannot be deserved and just. When I wrote this paper, I believed I was the first to think of this radical view. I later discovered that Henry Sidgwick had anticipated it in *Methods of Ethics*. I confess to having read Sidgwick's book years before writing my paper, but at the time of writing it I did not recall the relevant bit and, so, failed to refer to it. In contrast to me, however, Sidgwick does not endorse this radical egalitarianism. His verdict is that it brings us "to such a precipice of paradox that Common Sense is likely to abandon it" (*The Methods of Ethics*, p. 284). It is true that this radical egalitarianism is so far removed from common sense morality that it will be difficult to persuade the general public to accept it. Nevertheless, I want to insist in opposition to moral intuitionists that there is nothing sacrosanct about common sense morality (for reasons that will surface in my reply to the next question) and, so, the fact that a moral view conflicts head-on with this morality does not automatically disqualify it.

It might however seem that a radical, interspecies egalitarianism not only deviates from common sense, but is patently absurd. Is it not absurd to hold that we have reason to see to it that all non-human animals — be they hedgehogs, blackbirds, swordfish or whatever — become as well off as human beings? Nevertheless, here we must remember that such a principle of equality cannot be the only moral principle. As has been remarked, reasons of beneficence, in a revised shape, also survive the critique of common sense morality. If there are no general rights, these reasons are not counteracted by stronger reasons of rights. Moreover, my claim that considerations of personal identity are rationally irrelevant supports an imposition of a *requirement of universalizability* upon reasons of beneficence. Taken together this points in the direction of a sort of utilitarian morality with egalitarian distributive constraints.

However, it should be a utilitarian morality which distinguishes between "higher" and "lower" forms of utility or well-being, for reasons I delineate in "The Root of the Repugnant Conclusion and Its Rebuttal". With this adjustment we can see why we are not morally required to engage in the enterprise of trying to raise hedgehogs, etc. to our level of well-being. This would be futile, for higher quality well-being is likely to remain the unique asset of human beings. A weighing of utilitarian and egalitarian considerations against each other then yields the result that we ought not to equalize well-being if doing so would decrease the total sum or quality of well-being too much. The weighing of these principles against each other is however an imprecise affair which is unlikely to command consensus. But, in any event, it seems indisputable that this revisionist morality is more demanding on us than common sense morality. For instance, it doubtless requires those of us who are fortunate enough to occupy the affluent part of the world to aid people in the developing world to a much greater extent than we actually do.

This will have to suffice as an illustration of how I think moral philosophers should do normative ethics. The first step is to make an inventory of the content of common sense morality. The next step is a test of the validity of this content which takes us into branches of philosophy other than ethics, e.g., the metaphysics of free will and personal identity. The outcome will be a thoroughgoing revision of common sense morality, of the kind I have roughly sketched.

**How do studies within scientific disciplines contribute to the development of normative ethics?**

I believe that the relatively new discipline of *evolutionary psychology* can help us to understand why we have the kind of common sense morality that we actually have. It is natural to construe the general rights to ourselves and our property that this morality attributes to us as a development out of the special fierceness with which non-human animals defend themselves, the prey they have hunted down, their nests and so on. These dispositions being of obvious survival value. Moreover, a rights-based morality, which requires only moderate aid to others may well be conducive to the survival of groups in which it reigns. Similarly, desert-thinking can be seen as having developed out of a "tit-for-tat" practice which has been found to be beneficial for populations in which it is widely spread. "Tit-for-tat" consists in returning favours with favours and aggression with retaliation. It is easy to interpret such responses as being deserved, e.g., when we are angry, we naturally think "I'll give him what he deserves!".

Along such lines it may be possible to explain why we have the common sense morality we actually have without assuming its correctness: in rough outline it is found in human societies all over the world because it has had survival value, and not necessarily because it reflects a perceived moral truth. Thus, it could legitimately be exposed to revisionist criticism. A defensible morality must stand up against a backdrop of the extensive empirical and philosophical knowledge we currently possess.

## What do you consider the most neglected topics and/or contributions in normative ethics?

I have suggested a methodology for the philosophical discipline of normative ethics which has one descriptive and one justificatory part. (1) The descriptive part unravels common sense morality, taking pains to ensure that it appears intelligible in the light of what we know about human nature. (2) The justificatory part tests the validity of this morality in the light of our empirical and philosophical knowledge and, if called for, proposes revisions that pass this justificatory test. If we follow this methodology, the normative principles we end up with will have both an explanatory background in our nature and a justification upon the basis of our knowledge of ourselves and our environment.

It seems to me that moral philosophers sometimes launch fundamental normative principles that do not comply with these methodological demands. For instance, a rival to the sort of egalitarian view that I have adumbrated is *prioritarianism*, according to which benefits to those who are worse off *absolutely* – not in relation to someone else – have a greater moral weight or value than benefits to those who are better off (see Parfit, "Equality or Priority?"). Many people prefer prioritarianism to egalitarianism because they suppose that the former is not vulnerable

to the so-called *levelling down objection*, to which egalitarianism
falls victim. This objection is to the effect that egalitarians are
committed to holding that a levelling down which consists in the
better-off losing some of their well-being and sinking down towards
the worse-off is in *one* respect for the better, since there is now
less inequality, though it is better *for* nobody. This implication is
supposedly counter-intuitive.

As I argue in "Equality, Priority and Person-Affecting Value",
prioritarians are as much as egalitarians committed to there being
an *impersonal* value, a value that is not a value *for* somebody, as
the benefits distributed are (e.g., pleasure is good *for* the subject
who is feeling it). For prioritarians hold that it is (morally) *better*
that a benefit goes to a recipient who is worse off than to a recipient who is better off, though the benefit is, definitionally, as good
*for* the latter as for the former because it is the same benefit. But
then a levelling down could be an improvement in respect of this
impersonal value, for after a levelling down the (smaller) amount
of benefits distributed are in hands where on average they have
a greater moral weight or value. If so, prioritarians cannot wield
the levelling down objection against egalitarians. This objection,
if sound (which, following, e.g., Larry Temkin, *Inequality*, chap.
9, I would dispute), would then tell against both egalitarians and
prioritarians.

The reason why prioritarians have not seen that their view may
be exposed to the levelling down objection is probably that they
have not realized that they are committed to an impersonal, moral
value, along with the personal value of benefits. Consequently,
they have not understood that they are committed to answering
such questions as whether or not a change with respect to this
impersonal value, which takes place when there is a levelling down,
is for the better. Thus, prioritarianism is crucially underspecified,
and it is arguable that, if we try to work it out, it will be vulnerable
to the levelling down objection, or even worse objections.

Such underspecification is not an unusual defect of pure philosophical inventions conceived to mend some (alleged) shortcoming
in the common sensical conceptual scheme. (A similar case in point
is Samuel Scheffler's "agent-centred prerogatives" or options to
favour ourselves; see his *The Rejection of Consequentialism*.) In
contrast, I have tried to indicate how egalitarian considerations
arise out of a familiar, age-old concern with justice. There is nothing to correspond to this sort of background in pre-reflective, pri-
mordial moral dispositions in the case of the prioritarian weight.

But my feeling is that we get the wisest fundamental normative principles by a rational refinement of our pre-reflective moral intuitions. These principles will combine, so far as it is consistent, the virtues of the test of time and of reason. Therefore, when presented with putative fundamental norms, we should look around for factual backgrounds which could both explain our spontaneous adherence to them, or approximations to them, and justify them, or refinements of them. As I have indicated, these refinements may take us a long way indeed from the original.

## What are the most important problems in normative ethics and what are the prospects for progress?

I have outlined a revisionist morality which strongly diverges from common sense morality, and which is considerably more demanding. Against such types of moralityBernard Williams has objected in "Persons, Character and Morality" that we cannot reasonably be morally required to give up our "ground projects", the pursuits of which make our own lives meaningful. In reply to this objection, I contend in *The Retreat of Reason*, part IV, that morality should leave room for the pursuit of *ideals*. The idea is roughly that, if we are permitted to pursue something, although it goes against the goal of maximizing the fulfilment of *our own* lives, it must be permissible to pursue it, even though it goes against maximizing the fulfilment of *other* lives, since differences as regards personal identity are unimportant.

This reasoning presupposes that we have rejected the existence of rights that others could assert against us. But, apart from the fact that it may provide less moral room for autonomy than rights do, it appears to necessitate another sort of revision of common sense morality, a revision of its *grounding* rather than of its *content* which is what we have so far considered. As I have set out, this reasoning construes our moral reasons in an "internalist" fashion, as being dependent upon our desires. This yields a sort of moral relativism. It may however be a presupposition of common sense morality that, if we know the relevant empirical facts, represent them adequately to ourselves, and reason carefully, we shall all agree in our moral judgements. Perhaps this is because we then take it for granted that there are moral reasons which are independent of our attitudes, and which therefore are valid for all of us.

However, I cannot find any credible way of understanding such reasons which are external to our attitudes. Some, notably Parfit (personal communication), believe that truths about the existence of such reasons are necessary and known *a priori*, or through pure thinking, without any empirical observation. Now, we can no doubt know *a priori* that a proposition is necessarily true, or true in every possible world, if the proposition is analytic or true in virtue of its conceptual content. For thinking consists in relating concepts to each other. But propositions about what reasons we have are not supposed to be analytic or true in virtue of their content, and it seems to me a mystery how we could know *a priori*, or solely by thinking, that any synthetic propositions are true in every possible world.

Moreover, fundamental truths about external reasons do not satisfy a reasonable requirement on fundamental truths about reasons: that nobody could rationally deny their truth. For advocates of external reasons cannot plausibly claim that their philosophical adversaries without exception are irrational in denying the existence of such reasons. But imagine that it were possible for somebody to be aware of a fact, e.g., that X would involve great agony which, according to externalism, there is in fact sufficient reason to be intrinsically averse to, while rationally holding that there is no reason to be intrinsically averse to X (or anything). Then this person would be in the self-contradictory state of being both rationally required to be intrinsically averse to X – in virtue of awareness of the fact that X involves agony – and rationally permitted not to have this intrinsic aversion, in virtue of the rational normative belief held. Furthermore, if people could rationally disagree about fundamental truths about external reasons – both whether they exist and what they are about – one wonders how such disagreements could be rationally settled. There seems to be no way of doing so, since appeals neither to conceptual analysis nor to sense-experience will do.

Consequently, I think we may have to get by with moral reasons which are desire-dependent. This may necessitate another sort of revision of common sense morality, a *metaethical* revision of the *meaning* of our claims about what we have moral reasons to do, alongside the *normative ethical* revision about *what* we have moral reasons to do. So, when doing normative ethics philosophically, we may eventually find ourselves in the awkward position of having to offer to the general public a morality which both radically diverges from the common sense morality with which they have grown up, by being much more demanding, and which has a less authoritative grounding in their attitudes rather than in something external to these attitudes.

## Bibliography

Kamm, Frances, *Morality, Mortality*, volumes I & II (New York: Oxford University Press, 1993 & 1996).

Locke, John, *Two Treatises of Government* (London: J. M. Dent & Sons, 1990 [1690]).

Parfit, Derek, *Reasons and Persons* (Oxford: Clarendon Press, 1984).

Parfit, Derek, *Equality or Priority?* (The Lindley Lecture, The University of Kansas, 1995).

Persson, Ingmar, "A Basis for (Interspecies) Equality", in Paola Cavalieri & Peter Singer (eds.), *The Great Ape Project* (London: Fourth Estate, 1993).

Persson, Ingmar, "Genetic Therapy, Identity and Person-Regarding Reasons", *Bioethics* 9 (1995), 16–31.

Persson, Ingmar, "Person-Affecting Principles and Beyond", in Nick Fotion & Jan Heller (eds.), *Contingent Future Persons* (Dordrecht: Kluwer, 1997).

Persson, Ingmar, "Equality, Priority and Person-affecting Value", *Ethical Theory and Moral Practice*, 4 (2001): 23–39.

Persson, Ingmar, "The Root of the Repugnant Conclusion and Its Rebuttal", in Jesper Ryberg & Torbjörn Tännsjö (eds.), *The Repugnant Conclusion* (Dordrecht: Kluwer, 2004).

Persson, Ingmar, *The Retreat of Reason — A Dilemma in the Philosophy of Life* (Oxford: Clarendon Press, 2005).

Persson, Ingmar, "A Defence of Extreme Egalitarianism", in Nils Holtug & Kasper Lippert-Rasmussen (eds.), *Egalitarianism: New Essays on the Nature and Value of Equality* (Oxford: Clarendon Press, 2006).

Persson, Ingmar, "Rights and the Asymmetry Between Causing Bad and Good Lives", in Melinda Roberts & David Wasserman (eds.), *The Non-Identity Problem* (Dordrecht: Kluwer, forthcoming).

Persson, Ingmar, "The Act-Omission Doctrine and Negative Rights," *The Journal of Value Inquiry*, forthcoming.

Scheffler, Samuel, *The Rejection of Consequentialism* (Oxford: Clarendon Press, 1982).

Sidgwick, Henry, *The Methods of Ethics*, $7^{th}$ ed. (Indianapolis: Hackett, 1981 [1907]).

Spinoza, Baruch, *Ethics* (New York & London: Hafner, 1949 [1675]).

Temkin, Larry, *Inequality* (New York: Oxford University Press, 1993).

Williams, Bernard, "Persons, Character and Morality", in Bernard Williams, *Moral Luck* (Cambridge: Cambridge University Press, 1981).

# 11

# Janet Radcliffe Richards

## Distinguished Research Fellow

Uehiro Centre for Practical Ethics

Faculty of Philsophy, University of Oxford, UK

---

**Why were you initially drawn to normative ethics?**

In one sense there wasn't really a question of being drawn; I was
brought up to take ethics seriously. My father was a Unitarian
minister, and moral principles go naturally with a religious upbringing, even – perhaps especially – at the heretical margins of
Christianity. I remember deciding on my twelfth birthday that I
was now fully responsible for my actions, and soon after that –
or perhaps before; I can't remember the exact timing – I became
a vegetarian. I didn't know any vegetarians at the time – there
weren't many then – but I do remember trying to convince myself
at school dinner that there was no moral difference between the
corned beef and cabbage on my plate. I failed, and went home to
announce that I was going to be a vegetarian. My long-suffering
mother put up with it without a murmur; my father was pleased
that I was developing principles.

I suppose that means I started thinking about normative ethics
quite early; but it was a long time before I started working on
it as an academic subject. I discovered philosophy as an undergraduate, and that snuffed out the last flicker of religious belief,
but that had no effect on my views about ethics. Unitarians were
always pretty close to humanists in such matters (as evidenced by
the relationship of Harriet Taylor, who was a Unitarian, and John
Stuart Mill). There was very little ethics in our course, and anyway philosophical ethics at the time was of not much normative
use: those were the days of Moore, Hare, Ayer and Wittgenstein.
Somewhere along the line we picked up the basics of utilitarianism,
and somehow I began to realize that the moral impulse underlying
my vegetarianism fell a long way short of a fully worked out moral

theory, but that was about all. When after some years I went back to graduate work, and eventually on to university teaching, it was mainly in metaphysics and philosophy of science.

The shift towards normative ethics as a subject came purely by accident, when I was approached out of the blue, in the mid-Seventies, about writing a philosophical book on feminism. As I had no particular interest in feminism at the time, and no background at all in the relevant parts of philosophy, I still suspect this approach may have been a case of mistaken identity. At any rate, when I arrived to meet the Routledge editor to whom I had apparently been recommended, I turned out not to be the person he was expecting. He had no idea whether he had the name right and the face wrong or the other way round; and as he could not remember who had made the recommendation – some passing person at a philosophy conference – neither of us ever found out who had really been intended. Still, the challenge was irresistible. I wrote a proposal and was offered a contract, and *The Sceptical Feminist*[1] followed some four years later.

Since the project was undertaken completely from scratch, with no obviously relevant philosophical knowledge in the background, it was essentially a matter of trying to apply philosophical thinking directly to a practical subject; and this turned out to provide – for me, at least – an unfamiliar and enlightening direction of approach to normative ethics. Starting with practical problems and progressing to theoretical questions about what was needed to address them, rather than starting with normative principles and trying to apply them to practice, meant that the project worked not only as a substantive enquiry into feminism itself, but also as a series of experiments in the methodology of reasoning in normative and practical ethics.

When I began, I think I took it for granted that applied ethics would be much easier than the more abstract parts of philosophy, and that when the feminism job was done I would return to the simmering work in metaphysics. But the book was eventually published only because it became clear that the publisher would not wait much longer, not because the work was anywhere near finished. Applied ethics turned out to be in some ways even more difficult than abstract philosophy. The problems were in them-

---

[1] Janet Radcliffe Richards, *The Sceptical Feminist: a philosophical enquiry*; Routledge and Kegan Paul, London and Boston, 1990; Pelican Books, 1982. Reprinted with additional material, Penguin Books 1994.

selves just as complicated, and there were the additional demands of working out how to make all the links with familiar, every-day discussions. It does not matter much if the higher levels of philosophical logic or philosophy of physics are incomprehensible by the uninitiated or uninterested reader, but it does matter if philosophical progress in moral thinking cannot be clearly linked to current debates.

In fact the work in practical ethics became harder as it went along. There is still no end in sight, and the prospects of ever resuscitating the metaphysics are (alas) now so faint that I have almost forgotten them.

**What example(s) from your work (or the work of others) illustrates the role that normative ethics ought to play in moral philosophy?**

People who take normative ethics seriously are usually ones whose ultimate concern is practical. They think it matters how people live their lives, and are hoping to establish a set of standards by which to assess the choices available to them. But the further you go into the problems the harder they become; and it sometimes looks as though the earth may be swallowed by the sun before many philosophers reach conclusions that satisfy even themselves, let alone others. One of the most interesting problems in normative ethics, therefore, is to find ways in which philosophical enquiry can be of practical use before it achieves its ultimate goal.

This was, in retrospect, partly why it was so interesting to approach ground-level questions about feminism and politics di-rectly, rather than by way of theoretical investigations of norma-tive ethics. Analysis of practical issues is usually called applied ethics, which carries the top-down implication of starting with a set of standards and applying it to particular issues. But what working on *The Sceptical Feminist* began to suggest was that even while a fully-fledged normative theory remained out of sight, very small steps close to the ground could be recognized as making real progress. At any rate, I found that working through the logic of arguments – quite irrespective of any new empirical evidence – changed my preliminary ideas about at least half the subjects I tackled. That certainly felt like progress.

Although it was not obvious at the time, it became clear later that patterns were emerging in the most interesting lines of argu-ment. These were extremely simple, and in retrospect so obvious that they seem hardly worth mentioning. Still, what looks obvious afterwards often takes an extraordinarily long time to pin down in the first place, and making these techniques explicit has turned out to be surprisingly enlighning – again, to me at least – in all kinds of contexts.

The most straightforward of these techniques was simply checking that an argument offered in support of some policy or action really did entail its conclusion, quite irrespective of the merits of either premises or conclusion. If it did not, it was clear that the defender of the position needed to find another justification; and if the premises and conclusion were actually in conflict – as happened surprisingly often – then either the conclusion or the justifying argument needed to be rejected.

So, for instance, when the so-called Woman Question became a matter of familiar debate in the nineteenth century, traditionalists defended the laws and conventions keeping men and women in their separate spheres by insisting that these were justified because the sexes were different by nature, and therefore suited to different activities. But, as Mill demonstrated at the time[2], whether or not the claims about sex differences were true,[3] the usual arguments in defence of sex-differentiating laws and conventions simply did not work. It was claimed that women were excluded from the professions and political participation because they were incapable of doing mens' work, but, as Mill pointed out 'What women by nature cannot do, it is quite superfluous to forbid them from doing; what they can do, but not so well as the men who are their competitors, competition suffices to exclude them from"[4]. It was also claimed that the justification for women's legal subordination to their husbands was women's own preference for that position. But, if women really had such preferences, Mill said, all the laws and conventions designed to keep them in their place would have no purpose. Women would choose that place anyway, without any need for compulsion.[5]

In other words, if women were indeed as traditionalists claimed

---

[2] John Stuart Mill, *The Subjection of Women*, 1869; reprinted in John Stuart Mill, *On Liberty and Other Essays*, World's Classics, Oxford University Press, 1991.

[3] ".....a subject on which it is impossible in the present state of society to obtain complete and correct knowledge — while almost everybody dogmatizes upon it......" *ibid* p.494

[4] Ibid. p 499

[5] *Ibid.* p 499

they were, the sex-differentiating rules would have nothing do because the sexes would end up in their separate spheres even if subjected to the same rules. Conversely, if the differentiating rules did achieve anything, it must be to keep women out of work they could do and force them into relationships with men they would not have chosen. Either way, the prevailing liberal principle that people should be allowed to achieve whatever their abilities and inclinations allowed could not justify sex differentiation in law. On the basis of their own claimed premises, Mill's opponents should have reached his conclusion, rather than their own.

I said that such failures of argument were obvious when pointed out, and they are; but the reason why they need pointing out is that when people are resisting an opponent's conclusion, they almost always do this by attacking the argument's premises. If your opponents, who want to keep men and women in their traditional places, start by saying that men and women are quite different by nature – and especially if those differences amount to scarcely disguised female inferiority – it is hardly surprising if your immediate response is to deny outright that there are any such differences. But there are two real dangers in doing this. The first is that you make things too easy for your opponents, by missing the opportunity to drive them into a logical corner. The second, which is much more serious, is that you yourself implicitly accept your opponents' implication that the case turns on the question of whether or not such differences exist, and perpetuate their mistake.

I don't know to what extent these oversights actually do account for the insistence, still almost universal in academic feminism, that differences of character between the sexes are socially constructed. My impression is that it is at least part of the reason; and it also seems to me, for what it is worth (not much, probably, without the supporting argument) that it has been a disastrous mistake for feminism, entrenching mistaken traditionalist assumptions, and in effect closing enquiry into what should be one of feminism's main concerns. I also suspect that a similar mistake may underpin another feminist enterprise, equally widespread and in my view even more radically mistaken, of seeking and advocating 'feminist' approaches to the fundamentals of ethics, science, epistemology and many other areas of enquiry[6]. That, however, needs detailed

---

[6] See Janet Radcliffe Richards, 'Why feminist epistemology isn't', in *The Flight from Science and Reason*, eds. Paul R Gross, Norman Levitt, Martin W Lewis, Johns Hopkins, 1997; reprinted with additional material: 'Why feminist epistemology isn't, and the implications for feminist jurisprudence"

argument, not just assertion.

Of course the effectiveness of purely logical analysis of political and moral arguments depends on there being logical mistakes in the first place, but in fact the phenomenon is surprisingly common. All people, including all philosophers, hold masses of individual beliefs and convictions that are in tension or even incompatible with each other, and we seem as a species to be adept at devising spurious connections to justify our inclination to hang on to them all at once. In the nineteenth century many people who accepted, in principle, the liberal ideal of allowing people to find their own place in society nevertheless also held deeply ingrained traditional ideas about the appropriate relationship of the sexes, and were so convinced of the truth of both that they managed not to notice the flagrant fallacies involved in trying to derive the second from the first. (And, it should perhaps be added; they held both convictions so strongly that Mill's arguments had little practical effect at the time.) Similar devices for holding together incompatible convictions are to be found everywhere. In bioethics, for instance, I think it can be shown that the same thing happens when people try to justify deep feelings about the wrongness of a wide range of practices (euthanasia[7], prostitution, selling body parts, genetic engineering, and many others) in terms of the values they unhesitatingly profess in other contexts.

Furthermore, the value of this approach does not lie just in the opportunity to trounce confused opponents. It is in essence simply a method of enquiry that depends on seeking out incompatibilities in sets of beliefs; and since we all spend much of our lives trying to prevent incoherent systems from collapsing around us, it can be disconcertingly enlightening.

This also applies to the second technique. It is an equally innocent and intrinsically uncontentious variation on the first, even closer to the ground and even less commonly recognized in familiar debate, which has turned out to be even more productive. It consists simply of taking some defended or recommended pol-

---

*Legal Theory*, Volume 1, no. 4, December 1995, pp 483-518. The mistake is only in applying feminist enquiry to the fundamentals of ethics; there are plenty of feminist enquiries to be made at less deep levels.

[7] See e.g. Janet Radcliffe Richards 'Euthanasia', *Nature Medicine*, July (pp 618-620); and 'Thinking straight and dying well', Paterson Lecture, Voluntary Euthanasia Society of Scotland 1994; reprinted Sylvan Barnet and Hugo Bedau (eds) *Current Issues and Enduring Questions: a guide to critical thinking and argument*, Bedford Books, November 1995.

icy – typically, but not necessarily, one that you are inclined to challenge – and asking what principle could justify it. This can be done either after some purported justification has failed, as above, or before one has even been offered.

This was an approach I first stumbled on while trying to get to grips with the abortion debate. It seemed to me at first that abortion was simply not a feminist issue. If feminism is of any moral interest (rather than just a pursuit of women's interests) it must be about justice; and as a matter of logic[8] you need your theory of justice before you can say women are wronged by its standards. If the unborn child has a moral right to life then prohibition of abortion is *not* unjust, and to claim that there can be no such right because this would harm women is straightforwardly question begging. The important question is one that can be settled, if at all, only at the most fundamental level of normative ethics.

But in fact a quite different line of enquiry did turn out to be relevant to feminism. Many people intuitively support the idea (actually enshrined in some abortion legislation, including that of the UK) that abortion is generally wrong and should not be allowed on demand, but that it is permissible under certain specified circumstances: in case of rape, for example, or serious foetal abnormality, or danger to the life or health of mother[9]. This may sound like a good, moderate, commonsensical kind of policy; but if we are trying to decide whether to accept it we need to ask what moral principle provides its underpinning – partly so that we can assess that principle and partly so that we can follow it in other contexts if it does not turn out to be acceptable.

Once again, this simple approach has the great advantage of sidestepping the intractable controversies that are usually the first to arise in familiar abortion debates. It does not raise moral questions about the status of the foetus, or empirical ones about effects of abortion on mothers. The challenge is not yet to find an *acceptable* underpinning principle for the recommended policy but, much more humbly, *any* principle that will do the logical job. It may seem that this is a simple matter, and that the difficulties will not appear until the next stage, of assessing competing principles.

---

[8] Not necessarily chronology; as a matter of procedure you may well start with an intuition that women are unjustly treated. But the eventual demonstration of this will require a principle of justice from which the conclusion is derived.

[9] Current UK abortion law technically keeps abortion as a prima facie criminal offence, while allowing a wide range of defences.

However, this turned out not to be the case; and this was one of the contexts where the results of the investigation really took me by surprise.

The details take some explaining, and since the argument is heavily dependent on them it is frustrating not to be able to describe them here[10]. Essentially, the problem is that although it is easy enough to think of justifications for allowing abortion in the special cases, what is needed is a principle that will defend *both* those exceptions *and* the prohibition of abortion in general. The principle that life is sacrosanct will not do the job, because that would not allow the exceptions either. And, preposterous as it will undoubtedly sound, there was one principle I could find that came anywhere fulfilling the logical requirements. It was that errant women should be punished by being forced to bear unwanted children while the innocent were protected from harm—where the errant understood as the women who have had sex without specifically wanting children, and the innocent women who either did not want sex (were raped) or who did want children but through no fault of their own turned out to be carrying a defective foetus, or one that threatened their own life or health.

The fact that such an outlandish principle would provide a more or less coherent justification does not in itself imply that it is the *explanation* for the fact that so many people seem to find the view so intuitively attractive. Its purpose in the argument is just to sharpen the original challenge: if you recommend a policy of this kind, but are not willing to accept this remarkable principle as the underpinning justification, then you need either to find an acceptable principle that does the logical job, or to revise your policy. But of course it can be regarded as providing a *hypothesis* about motivation; and interestingly, other parts of the literature began to provide evidence that feelings along these lines did actually exist (for instance, denial of anaesthetics during abortion). This began to suggest the considerable extent to which the abortion issue could in practice be seen as genuinely feminist by almost any standards, irrespective of controversies about the value of human life.

Both elements of this technique – challenging defenders to produce a justifying principle, and in the meantime producing candidates that would do the logical job and at the same time provide hypotheses about real motives – have also turned out to be illu-

---

[10] *The Sceptical Feminist, op cit* 265 - 281

minating in other contexts. For instance, also within feminism, it could be applied to the problem of justifying the particular forms historically taken by the subjection of women to men[11]. The only principle that seemed to do the necessary logical work was one that depended on men's controlling women in ways that allowed them to identify and own their children – which, regarded as a hypothesis about causation, has also been increasingly supported by empirical work. But the potentiality of the approach is endless. It is quite extraordinary how often, when justifications for actions and policies are assessed in sufficiently close detail, the only ones that turn out to fit the logic are ones that their proponents would not want to accept as principles – at least in public – and also look disconcertingly plausible as accounts of motivation. For instance, it is an extremely revealing technique to apply in the context of clinical medical ethics. It can bring about significant changes of mind, easily recognized as real progress in moral thinking.

Finally, there eventually developed an even more powerful technique, a further extension of the other two, but starting to incorporate moral substance as well as logic. You take some policy your opponents are recommending, and instead of looking directly at their arguments for it, you show that if you take some principle those opponents themselves normally profess (and, ideally, that just about anyone would accept), it provides a prima facie case for the opposite conclusion. So, to take an issue in which I became accidentally embroiled some years ago and from which I still have not managed to escape, you look at something like the widespread insistence that organ selling should not be allowed, and show that uncontroversial moral principles – such as the intrinsic desirability of saving life, giving the poor as many opportunities as possible to improve their lives, and allowing adults to make exchanges they regard as mutually beneficial, all suggest a presumption in favour of *allowing* organ selling.

Of course this does not make the case, because a presumption in favour of some policy – which is all the argument demonstrates so far – is potentially defeasible. But it is a methodologically powerful beginning, because it shows that any argument produced to defend the prohibition of sales must be powerful enough not

---

[11] *Op cit* 176; also Janet Radcliffe Richards, "Separate Spheres" in Peter Singer (ed.) *Applied Ethics*, Oxford Readings in Philosophy, Oxford University Press, 1986 pp. 185-214; "Metaphysics for the marriage debate", *San Diego Law Review* 2005, vol. 42, issue 3, p. 1125-1142

only to work in its own right, but also to defeat the presumption against prohibition. As you would expect in a context that arouses such strong feelings, these attempts stretch in never-ending line; but I have yet to see one that does not involve corkscrew logic or flagrant invention of facts. I don't like the idea of organ selling any more than anyone else does; but on the other hand, my likes and dislikes seem hardly relevant, given the strength of the presumption on the other side.

This is probably not the most persuasive illustration of the methodological point I am making, since anyone who is sure the conclusion is outrageous will presume there must be something wrong with any arguments that suggest otherwise. I can only suggest they look at the details[12]. To me it seems that uneasiness about the conclusion is part of the evidence that this kind of technique really can make moral progress. And this one, too, can be used in innumerable ordinary contexts.

### How do studies within scientific disciplines contribute to the development of normative ethics?

Consistency is a necessary condition of a satisfactory normative theory, but it is obviously not sufficient. The discovery of inconsistency does not, in itself, determine the direction in which it should be resolved, and, although the possibilities turn out to be much more limited in practice than in theory, there is still a lot of scope for real disagreement. In the abortion issue, for instance, people who decide they must reject the apparently moderate policy discussed earlier might move to a principle that demanded total prohibition, or one that allowed abortion on demand, or some kind of time-based principle. So the question arises of what other means might be employed to eliminate candidate normative theories that had survived the consistency test. This is, I think, where the relevance of science comes. For me personally this started to become clear many years later, through work on the implications of Darwinian theory. This was another project taken up from scratch,

---

[12] See e.g. Janet Radcliffe Richards, 'Nephrarious goings on: kidney sales and moral arguments', *Journal of Medicine* and *Philosophy*, Vol 21 no 4, August 1996, pp 375–416; Janet Radcliffe Richards et al, 'The case for allowing Kidney Sales', *Lancet* 1998; 351: 1950–52; Janet Radcliffe Richards, "Organ trade: the philosopher" Ch. 37 in *Living Donor Organ Transplantation* eds. Rainer W.G. Gruessner and Enrico Benedetti, McGraw-Hill Professional Publishing, New York (in press).

more or less by accident, at someone else's suggestion[13]; and, like the work on feminism, it opened up what were for me new directions of approach to questions of normative ethics.

Here are three that particularly struck me, necessarily more asserted than argued for.

## Changing world views

One effect of the advance of science has been its bringing into the domain of nature phenomena that had originally seemed to demand supernatural explanation. In this way it has gradually eroded many of the original reasons for believing in God; and the tipping point of this process was, for many people, the Darwinian explanation of how natural order and complexity could evolve without any need for a designer.

The obvious relevance of this for normative ethics is that many traditional ethical views make little or no sense without an underpinning religion. If God doesn't exist you can't reasonably direct your life to pleasing him, or think that all will eventually be well if you follow divine commands. Without religion, Cardinal Newman's well-known exposition of the Catholic view, for instance, that "it were better for sun and moon to drop from heaven, for the earth to fall, and for the many millions upon it to die of starvation in extremists agony .... than that one soul ....should commit one single venial sin ...."[14], makes no sense at all. Even though most secularists would probably accept the prima facie wrongness of the lying and stealing Newman gives as examples of venial sins, it is hard to imagine their regarding the avoidance of one such action as more important than preventing agony for millions.

Of course Newman's view is extreme even in religious terms. Nevertheless, there are real differences between secular and religious views of the world, and it seems clear that at least part of the reason why many of our moral beliefs are confused is that Western and Near Eastern ideas of ethics have deep roots in the Abramic religious traditions, and persist even when those traditions have been rejected or radically modified. On the one hand people who have in theory abandoned religious belief still carry many of its presuppositions, and, on the other, the kind of thor-

---

[13] Janet Radcliffe Richards, *Human Nature After Darwin: a philosophical introduction,* The Open University (Milton Keynes) 1999; substantially revised edition Routledge (London and New York) November 2000.

[14] John Henry Cardinal Newman, *Apologia pro Vita Sua,* Chapter 3.

oughgoing theology embraced by Newman has to a large extent had its edges rubbed off by liberal secularism. The result is a persistent attempt by members of both groups to rationalize deeply held intuitions in essentially secular terms, and the kind of incoherence that provides so much scope for purely logical analysis. When the inconsistencies are clearly identified, religious believers may reject what they regard as secular contamination; but people who have rejected religion may recognize the residual religious presuppositions in some of their intuitions and decide to achieve consistency by abandoning those.

There is also a wider, more subtle, and less well understood way in which a proper understanding of the implications of Darwinian theory undermines much traditional ethical thinking. It is not merely that it removes one of the main reasons for belief in God. Properly understood, it also implies a total split between the natural and any kind of moral order. Much supposedly nonreligious thinking involves quasi-religious assumptions that the world has an underlying moral order, and that things go wrong only because we are acting wrongly. This appears, for instance, in large tracts of current ecological thinking; and my own view is that something like it is needed to make sense of currently popular ideas of virtue ethics.

It seems to me—though obviously this needs argument that once you have accepted a Darwinian explanation of the existence of natural order, and fully recognized the total separation of the natural and the good, the effect on ethics is radical. As regards fundamentals, ethics must ultimately be concerned with outcomes, rather than rules or characters. There is plenty of space for *derivative* deontological rules and for the cultivation of virtues, but the form those rules and virtues should take must itself be analysed in broadly consequentialist terms. That has considerable implications for normative ethics.

### Science in consequentialism

To the extent that normative ethics is concerned with outcomes, it of course involves science in another way – a more local way – in that making things better depends on knowing how they work. This was what Bentham meant by his claim that he had turned ethics into a science. He was wrong, of course, since there remain the irreducibly ethical problems of working out the criteria by which to measure improvement; but putting any consequentialist view into practice does depend critically on the facts, and knowledge of the facts develops with the advance of science.

Here again, the Darwinian developments are particularly interesting. As Bentham and Mill realized, the development of the human and social sciences was going to be of huge importance in implementing utilitarianism of any form, and my own view is that the application of evolutionary thinking to these areas is leading to the most exciting progress there has ever been. It is true that the resulting theories of evolutionary psychology, anthropology and the like are still strongly resisted in many quarters; but although there are of course genuine scientific debates in this new area, it is clear that many of the objections are to the whole area of enquiry. My view is that that this resistance comes largely from serious misunderstandings of Darwinian theory itself and from mistaken beliefs about its political implications[15]. Getting these issues clarified in academic and everyday debate is going to be enormously important for the development of practical ethics.

## Human nature and debunking explanations

Finally, there are also interesting ways in which increasing scientific understanding of human nature and human history are relevant not just to the application of normative theories to practice, but also to the assessment of the irreducibly ethical questions themselves. As has already been suggested in the context of religion, recognizing the origins of intuitions may help to decide the direction in which to resolve conflicts between them. This may also happen if we can recognize the evolutionary origins of strong intuitions that are difficult to endorse intellectually.

For instance, there is the familiar philosophical problem of retributivist ideas of punishment. Most of us have a deep feeling that if people act badly enough they really deserve to suffer, and that inflicting suffering is intrinsically appropriate even though it may have a real social cost. In the strongest versions of these theories rewards and punishments are so thoroughly deserved as to be extreme and eternal. But many philosophers see insuperable problems about the kind of free will needed to underpin such a strong idea of desert. Either, everything is caused by what went before, in which case we not responsible because the causal chain was

---

[15] For instance, the widespread and totally wrong idea that ideas of Darwinian evolution provide any justification whatever for the idea of the disastrously misnamed ideas of Social Darwinism; see *Human Nature after Darwin*, Routledge, *op.cit* p.248, and 242-252 *passim*.

set in motion before we even existed, or some things are actually uncaused, in which case nobody and nothing could be responsible for them. Either way, ultimate responsibility is impossible. That doesn't imply that punishment must be unjustified, but it does imply that any justification must be consequentialist, essentially as a means to deterring greater harm. But this seems to have implications that most people find repugnant, such as letting off the guilty when punishment would serve no deterrent aim.

In this conflict of intuitions, which element should be regarded as the counterexample, showing the other to be wrong? Some philosophers think that when we have such deep feelings about the essential connection of suffering and desert we should go by those feelings, and presume that something has gone wrong in our arguments about free will. But your attitude to deep feelings may depend on your views about their origins. If you think they are rooted in a divinely implanted conscience, it may be reasonable to accept them as your fixed point (though there are always theological problems about distinguishing the word of God from the promptings of the devil). But in a Darwinian world you look for a different kind of explanation, and try to understand how these feelings might have been reproductively advantageous during the course of evolution. If you can account for the existence of strong feelings in this kind of way, you need no longer see their strength as giving any reason for moral endorsement of them.

That does not mean the feelings are irrelevant to moral reasoning. Any planning for a better society has to take account of people as they are; and if feelings of strong anger for injury are deeply ingrained instincts, there may be little point – in the present state of technology, at least – in trying to get them out of our system. But it makes a great deal of difference if such information can be recognized as belonging not to the deepest part of the normative enquiry (deciding which values to reject in case of conflict), but to our knowledge of the workings of the world. It can be demoted, as it were, from this section to the previous one: relevant to the application of moral principles, but not to moral reasoning itself.

**What do you consider the most neglected topics and/or contributions in normative ethics?**

The topics I am most concerned with aren't exactly neglected; it's just that they are still relatively undeveloped because they are so intimately concerned with the rapid development of science. There is a great deal of excitement about these new areas, but work on their implications has not been going for long. In particular, there needs to be far more interactive work between scientists and philosophers. It doesn't surprise philosophers that philosophically untrained scientists mistake their own conclusions for answers to philosophical questions; but at the same time much philosophical ethics seems to go on as if the science had never happened, or is based on serious misunderstandings of it.

Many of the problems – such as the fascinating questions about responsibility raised by developments in neurology – are rather subtle, and only academically inclined people are likely to be aware of them. But there are also many that nobody can avoid, in technological advances produce new situations and new choices. These are the ones the public hears about every day: decisions that have to be made about the use of reproductive technologies, genetic engineering, transplantation, treatments that draw out and separate the different elements of the dying process, and innumerable others. There is of course nothing neglected about these topics as such. Books and journals on bioethics and medicine are bursting with discussions of them, and everyone is familiar with the difference between arguments defending more or less conservative positions and the iconoclastic opposition. Nevertheless, most of these public debates are philosophically very weak. A great deal more sharpening of the arguments is needed to expose the underlying differences between irreconcilably different world views, and work out what to do about them. There are serious questions about how much practical agreement can be achieved against a background of radical moral disagreement, but we can't confront those questions seriously until we get rid of the fudged arguments that mask the differences.

There are also many issues that are undeveloped because they are difficult to confront for ideological reasons. This has, of course, been the case throughout history, and is inevitable in a social species. It takes heroism – or perversity – to ask questions that people around you think it is unacceptable even to raise. Questions about eugenics, human similarities, group equality, and population issues, for instance, are very difficult to discuss at present, but are too important to be disregarded or given only politically correct treatment.

## What are the most important problems in normative ethics and what are the prospects for progress?

Expressed in the most general terms, the most fundamental problems in ethics (metaethical as well as normative) involve sorting out the extraordinarily complex relationship between ethics as a rational enquiry and the facts about human nature – which seems to become increasingly difficult to unravel as we discover more about those facts. One increasingly important aspect of this is to understand the difference between how we work as individuals and small groups – the basis of our evolved common sense morality – and the totally different problems raised by the unintended consequences of billions of individual actions. We cannot imagine a morality that did not involve special concern for family and friends, but those are just the attitudes that lead to massive overpopulation, corruption and exploitation.

As for the prospects of progress – the first thing to be clear about is what counts as progress. That is another matter radically affected by the idea that the universe just is morally indifferent. A religious upbringing (I speak from experience) goes with the idea that there is a fundamental moral order, and that the reason why things go wrong is that people don't do as they should. Even if you don't understand the overall plan you can still make your contribution by doing what is right.

From this starting point, resignation to the recognition of Darwinian moral disorder is deeply disappointing. It doesn't in the least undercut the importance of ethics as such, and it doesn't make moral progress nonexistent, but the understanding of progress has to be understood very differently. In such a universe there is no point in trying to measure moral progress in terms of approach to some kind of utopia, partly because the kinds of improvement we can make are so minute, and partly because whatever situation we achieved, we would be able to imagine a better one.

Furthermore, even the purely philosophical project of trying to establish standards by which to assess whether things are improving or getting worse is affected. We can no longer expect to find clear instructions for leading our lives as they should be led. We can recognize various things as quite unequivocally good and bad in themselves (the badness of suffering is the obvious one), but many others are less clear and do not seem to be reducible to each other. And even if we could identify one fundamental value, questions about distribution are certainly not reducible to questions about what goods matter most. Incommensurability seems to lie at the root of ethics, as yet another aspect of the general messiness of things. And even if there were simple, clear answers at this level, there would still remain the endless problems of applying these standards to the elusive facts about the world, and devising standards to apply to everyday practice. It leads to an almost religious humility about our place in the scheme of things, but without the cosmological underpinning that supports religious hope.

That is why I think progress in ethics needs to be thought of less in terms of approach to some end than as rising from a starting point. We may not be able to envisage a perfect world, or even clear standards by which to measure overall progress, but there are some certainties we can hold on to. I still have no idea whether there is any real point in being a vegetarian, but there can be no doubt about some of the underlying impulses. For instance, suffering certainly matters; equal suffering matters equally wherever it is, whether in human, animal, Martian or even robot; and the imposition of great suffering in one place for minor gain in another is wrong. We can also recognize as objectively confused some arguments for resisting the good and condoning the bad. All that allows for some kinds of progress; and, as they say, it is better to light a candle than curse the darkness.

# 12

# Peter Singer

## Ira W. DeCamp Professor of Bioethics

Princeton University, USA

---

**Why were you initially drawn to normative ethics?**

As an undergraduate at the University of Melbourne, I became interested in philosophy, but I was also interested in social and political issues. It was the 1960s, the era of the Vietnam War and Australian troops – including conscripts – were serving there. Abortion law reform was another major question at the time. While it was fun puzzling over questions like "How can I be sure that I am not dreaming right now?" I couldn't really see this as something that was worth spending a lot of time on. It made no difference to the world, and when I tried to interest my friends in it, it struck most of them as silly. But ethics and political philosophy, connected with the social and political issues that interested me.

I was fortunate in having H.J. McCloskey as my lecturer in my first ethics course. He was then considered rather old-fashioned in his views, because instead of just doing meta-ethics, he devoted half of his course to normative ethics. He was a deontologist and an intuitionist, a follower of W.D. Ross, but he gave us a good introduction to utilitarianism and other normative theories as well.

Eventually ethics and political philosophy became the focus of my work. I did my Master's thesis on the topic "Why Should I Be Moral?" Then I went to Oxford to do a postgraduate degree, and also focused on ethics and political philosophy. I wrote my thesis on the obligation to obey the law in a democracy – an issue relevant to those who were opposing the war in Vietnam, some using lawful means while others used civil disobedience, and a few were prepared to resort to violence. I discussed the ethical issues around these choices. And around this time I was also challenged by some vegetarian fellow-students to think about the moral status of animals. A humanitarian crisis in what is now Bangladesh led me to write about the obligations of the rich to assist people in danger of starvation. So I started doing, not just normative ethics, but applied ethics.

Philosophy had for decades been dominated by the idea that the subject is all about the analysis of the meanings of words. In ethics the orthodoxy, as expressed by leading philosophers like A.J. Ayer, was that the field had nothing to say about what we ought to do. The student demand for "relevance" in university courses and the rise of a more radical younger generation of philosophers began to challenge that orthodoxy. *Philosophy and Public Affairs* was the first journal to publish scholarly articles in applied ethics, and my essay "Famine, Affluence and Morality" appeared in the second issue. Suddenly there were lots of exciting issues to explore.

## What example(s) from your work (or the work of others) illustrates the role that normative ethics ought to play in moral philosophy?

Normative ethics – and applied ethics, as a branch of normative ethics – ought to challenge us to think more deeply about what we ought to do. Three issues on which I've tried to do this are the ethics of how we treat animals, the sanctity of human life, and the obligations of the rich to those who are so poor they cannot satisfy their basic needs.

Some people are sceptical about the impact of moral argument on real life. They believe that moral argument is really a rationalization of what we wish to do, and rarely or never changes anyone's mind.[1] I believe all three of the issues I have mentioned offer counterexamples to this view, but the clearest one is the issue of the treatment of animals. As James Jasper and Dorothy Nelkin observed in *The Animal Rights Crusade: The Growth of a Moral Protest*, "Philosophers served as midwives of the animal rights movement in the late 1970s."[2] This movement has led to significant reforms in the ways in which experiments are performed on animals, and, especially in the European Union, to laws phasing out some of the worst forms of factory farming, including keeping veal calves and sows in crates so small that they cannot walk

---

[1] See, for example, See Richard A. Posner, *The Problematics of Moral and Legal Theory*, (1999).

[2] Free Press, New York, 1992, p.90.

or even turn around, and keeping hens in very small wire cages without any kind of nesting box to lay their eggs in, or enough room to perform basic instinctual behaviours. These reforms in the European Union will affect hundreds of millions of animals, and transform large industries – all because of an ethical concern for the welfare of animals. Now there are some hopeful signs that the United States is beginning to follow Europe's example. So here is an area of everyday life in which philosophy has played a truly critical role in society, not only at the level of ideas, but in instigating significant changes in society.

It is woth remarking that this modern philosophical challenge to the way we think about nonhuman animals came from those working in normative ethics in what is sometimes called the "analytic" tradition, that is, the tradition of English-language philosophy. Thinkers in the continental European tradition, the tradition of Heidegger, Foucault, Levinas, and Deleuze, played no role at all. Despite the much-vaunted "critical stance" that these thinkers are said to take to prevailing assumptions and social institutions, they failed to grapple with the issue of how we treat animals. Marxists also stood on the sidelines, or even jeered at the "radical chic" idea that animals are an oppressed group. It was the supposedly socially conservative analytic tradition that spawned this particular very radical critique of the status quo.

Why should this have been so? One reason may be that the British tradition of Hume, Bentham and Mill already had reached the conclusion that the capacity for experiencing pain and pleasure is what is crucial to moral status. In contrast, the continental tradition focused more on Kant and made the ability to reason and with it the capacity for autonomy, the crucial requirement. And on this point, at least, Marx remained typical of the Kantian tradition. He inverted Hegel's idealism, but did not question that it is the fate of human beings alone that really matters.

Still, it is astonishing that so few of Kant's followers noticed that his emphasis on reason and autonomy gave rise to a problem about the status of human infants and humans with profound intellectual disabilities. Clearly, if the ability to reason or to act autonomously, were what makes human beings "ends in themselves" rather than just the means to the ends of others, then some human beings would be just means to the ends of others, not ends in themselves.

Perhaps the deeper lesson to be learned from the failure of continental European philosophy to grapple with the issue of the moral status of animals is that to adopt a "critical stance" requires us to be critical about vague rhetorical formulations that appear profound or uplifting, but do more to camouflage weaknesses in reasoning than to hold these weakness up for critical scrutiny. Normative ethics should be less respectful of the authority of the "great" philosophers of the past, and more ready to punch a whole in inflated rhetoric that lacks clear argument – even if doing so makes us as unpopular as Socrates became when he did the same thing in ancient Athens.

## How do studies within scientific disciplines contribute to the development of normative ethics?

Some scientific disciplines contribute to normative ethics by raising new problems for ethicists to consider. This is particularly the case in bioethics. Advances in embryology and medicine made it possible to fertilize an embryo outside the human body and thus created the issue of the moral status of an embryo, independently of the question whether a woman has a right to control her own body. Being "pro-choice" was suddenly no longer enough to resolve issues about whether it is wrong to destroy an embryo. There are many other examples of new ethical issues generated by scientific developments. Issues of genetic selection are sharpened as our knowledge of genetics increases. Sex selection of embryos and fetuses has become possible, but should it be permitted? We are forced to make life-and-death decisions about people who formerly would have died irrespective of our best medical efforts, but who now survive in a condition that makes it doubtful if survival is a benefit to them or anyone else. Possibilities of enhancing human beings, either by genetic modification or by the use of drugs or other means are not far away, and we will soon need to be discussing them more seriously too.

Evolutionary biology, and more specifically evolutionary psychology, are relevant to normative ethics in a quite different way. They shed light on the origin of ethics, on where we get at least some of our normative judgments from. There now seems no doubt that some of our intuitive responses are not purely cultural – they have a biological basis. The neurosciences, and particularly brain imaging techniques, are also shedding light on what parts of the brain are involved in making moral judgments. Other forms of research in psychology and the social sciences add to the broader picture. I believe, as I've argued in more detail elsewhere, that our new knowledge in these sciences should make us more sceptical about our immediate, more or less automatic, intuitive responses to moral problems.[3]

## What do you consider the most neglected topics and/or contributions in normative ethics?

What we should be doing about population growth is somewhat neglected at the moment, since it does underlie many other issues, including poverty and climate change. Perhaps its relative current neglect is an overreaction to the overhyped writings about the "population bomb" and imminent mass famines that were popular in the 1970s.

I would also include the issue of what Nick Bostrom calls "existential risks" – how should we act in regard to risks, even very small ones, to the future existence of the entire human species? Arguably, all other issues pale into insignificance when we consider the risk of extinction of our species.[4]

But there aren't many seriously neglected topics. There are so many philosophers doing applied ethics nowadays that all the obvious gaps get filled by people trying to find something new to say. Granted, it's quite possible that in ten or twenty years we will discover some new topic of interest that is not discussed now, and wonder why philosophers of our generation neglected it. It's happened often before – for example, with the ethics of our treatment of animals, which was largely neglected for most of the last two thousand years—and no doubt it will happen again with some different topic. Perhaps we should soon start discussing the moral status of conscious machines.

As for neglected contributions, while the work of R.M. Hare is not entirely neglected, it is not now paid the attention it deserves. Compare the attention Rawls has received over the last 30 years – and yet Hare is, to my mind, a more rigorous philosopher. Mind you, I wouldn't want to see as much written about Hare as has been written about Rawls during those decades. That's excessive by any standards. So much discussion of any one philosopher becomes boring.

Going back further, I regret the fact that Mill's *Utilitarianism* is much more widely read than Sidgwick's *The Methods of Ethics*,

---

[3] 'Ethics and Intuitions', *The Journal of Ethics*, vol. 9, no. 3-4 (October 2005), pp. 331-352.

[4] See Nick Bostrom, "Existential Risks," http:// www.nickbostrom.com / existential / risks.html

despite the fact that *Utilitarianism* is a hastily-written work, full of doubtful arguments. *The Methods of Ethics,* which Sidgwick painstakingly revised 7 times over a thirty year period, is simply the best book on ethics ever written. It's difficult to think of any major issues in normative ethics that are not already touched upon there, and often it is hard to improve on what Sidgwick says. If students find it too long to read, then they should at least be referred to the last two chapters of Book III, all of Book IV, and the Concluding Chapter. But more people read Mill, no doubt in large part because Mill was the more concise and elegant writer.

### What are the most important problems in normative ethics and what are the prospects for progress?

In applied ethics, the most important problem is the obligations of the rich to the world's poorest people – especially in the light of climate change, which is largely caused by the rich but will have the most devastating impact on the poor. Climate change itself, of course, is a huge moral challenge, and it is fundamentally an issue of the just allocation of a scarce resource – the capacity of the atmosphere to deal with our waste gases.

Perhaps for the first time in human history, the rich have the capacity to eliminate poverty as a large-scale long-term phenomenon, reducing it to small pockets of poverty in isolated areas, or temporary emergencies as a result of war or climatic disasters. But is it also our obligation to do this? I believe it is, but that argument is not yet resolved. Although I am not naïve enough to believe that resolving it would be decisive in leading to action, it can contribute to change. Incidentally, slowing population growth is an important part of poverty reduction, but of course it raises some ethical issues of its own.

Issues relating to animals are also of great importance. There has been a lot of discussion of the use of animals in research but I think the most important ethical issue concerning animals is the way we use them for food, because the numbers dwarf all of our other uses of animals. We raise hundreds of billions of animals every year and capture and kill hundreds of billions of fish and other sentient aquatic creatures. Tens of billions of animals we raise are confined for their entire lives in factory farms. We are totally responsible for their existence, and it is usually a miserable one.

If we consider normative ethical theory rather than applied ethics, I would give prime importance to questions about the role reason plays in morality. There are two aspects of this question. One is whether we can say that one normative ethical theory is more rational than another – for example, can it be shown to be irrational to accept a deontological form of intuitionism, rather than a consequentialist view? The other aspect is the debate over what philosophers call internalism and externalism—whether to accept that you ought, morally, to do something also entails that you are irrational if you freely choose not to do it.

# 13

# John Skorupski

Professor of Moral Philosophy

University of St. Andrews, Scotland

---

**Why were you initially drawn to normative ethics?**

There is always something of a division between those drawn to philosophy by humanistic and normative – anthropocentric – interests and those drawn by scientific and metaphysical – cosmocentric – interests.

Take 'the nature of existence.' To anthropocentric philosophers this sounds an existential question about what it is to be human, how we humans stand in relation to the world; it may lead into questions about the world itself. To cosmocentric philosophers the nature of existence is in the first place a scientific question about what there is and how things work. Not that these two ways of being interested in the nature of existence are unconnected; on the contrary, science's often mentioned 'disenchantment of the world' assuredly has fundamental consequences for the existential question. Another topic that divides and yet necessarily brings into contact the two temperaments is the significance of scepticism. Cosmocentric philosophers tend to see scepticism as a distraction, or at best a purgative on the way to establishing sound method. Anthropocentric philosophers often see it as showing something significant about our relation to the world, the limits on our reason, or the limits of the scientific image.

My route into philosophy was definitely the anthropocentric one, and this remains my orientation. I see now that even as a student the questions that interested me were normative questions about meaning, justification, reasons. Over time my interests have also become increasingly meta-normative. Here too there is an anthropocentric drive; I want to know what normativity is as a part of knowing what being human is.

This, as much by accident as design, has led me in three directions. One leads into the heartland of philosophy. In recent decades the concept of a reason has come to seem to some philosophers to be central to an understanding of the normative domain. I agree with that, and am currently working on a book which will provide a unified treatment of reasons to believe, to feel and to act: hence (as I claim) of the three essential dimensions of normativity: the epistemic, evaluative and practical. The arguments of this book point towards a re-assertion of the Critical tradition in philosophy, i.e. towards a sort of qualified vindication of Kant's 'Copernican revolution'. (See 'Propositions about Reasons.')

A second direction leads into questions in ethics that cross borderlines between ethics, culture and politics. The essays in *Ethical Explorations* pursue this group of interests. The third direction leads into the history and evolution of what the French call *mentalités*: obviously a vast area within which one necessarily selects. I am an atheist but have always been interested in religion as a manifestation of human meaning-giving impulses. My first book, *Symbol and Theory*, was a study of theories of religion and magic in social anthropology – of the comparisons that have been made between those 'primitive' or 'elementary' world-views and ours, and of what those comparisons, made by moderns, show us about the modern outlook. For similar reasons I have been interested in $19^{th}$ century thought: it took a historical stance which I find congenial, it treated ethical, religious and political themes with unsurpassed imagination and human insight, at the same time it approached them with an admirably externalist explanatory interest.

Mill is the $19^{th}$ century thinker on whom I have spent most time. For various interesting historical reasons his philosophy has yet to recover the standing it deserves, though it is getting there. Recent work on him is beginning to show more interest in its overall metaphysical as well as its cultural and ethical shape. The same things, interestingly, are true of Hegel: the resurgence of interest in these two greatest moral philosophers of the $19^{th}$ century is curiously similar. Also interesting is what they have in common. Both tried to achieve an ambitious synthesis of enlightenment and romantic themes; both of them developed broad and layered ethical outlooks which the utopian simplicities, either/ors and voluntaristic illusions of modernism abruptly ended. Moral philosophy can benefit from an unprejudiced, fresh look at these two masters. (See *Why Read Mill Today?*)

**What example(s) from your work (or the work of others) illustrates the role that normative ethics ought to play in moral philosophy?**

There is no one role normative ethics *ought* to play; nor is there any one way of doing it. It should be the broadest possible activity in both matter and method. It can be closely tied to, or quite distant from, other forms of philosophising; it can be abstract and sweeping or concrete and specific, it can be historical, literary, analytic, etc.

That said, the way of doing it that I myself prefer links it to, rather than distancing it from, general philosophy on the one hand and history on the other. History, because ethical convictions develop in their own way through large social processes, with only indirect input from the ideas of philosophers. We can debate how much social relations and ideas influence each other; certainly ideas are effects as well as inputs. The philosophy of the normative domain deals with a living subject that strikes back.

Hence to me it seems that philosophical ethics must be as much concerned with critical interpretation as with advocacy. At its most general, it is an attempt to grasp the overall structure and force of our evaluative and practical thinking, taking into account all its obscurely related domains. (One should not isolate ethics: this is what ethics, aesthetics and politics taken together do. Their relationship is very much an issue that interests me, but too big to discuss here.) Among attempts at this kind of comprehension which I admire, without necessarily agreeing with them – they obviously don't agree with each other! – are Hegel's *Philosophy of Right*, many of Nietzsche's writings, Sidgwick's *Methods of Ethics* and many of his other writings. This for two reasons: the sense these philosophers have of the substantive 'thickness', possibly even incoherence, of the ethical, and their wish to engage with it in terms which can seriously mean something to us now (that can be 'actual'). They do not try to generate ethics as a whole from some simple single thesis, or to convince us that we should live by some anachronistic model – as virtuous Athenians, mediaeval fideists, optimistic *philosophes*, abstract lovers of humankind, etc.

As regards contemporary work I think I have specially learnt from that of Bernard Williams, Charles Taylor, and Tim Scanlon. Again they obviously don't agree with each other, and again I am speaking about what interests me, without meaning to suggest that these philosophers illustrate *the* way to do normative ethics. Nor do I agree with any of them overall. I am much more sympathetic to the utilitarian tradition than they are. I would say that Scanlon is the only one of them who appreciates how deeply one has to dig in any effort to stand up a radical alternative. (See his 'Contractualism and Utilitarianism', and *What We Owe to Each Other.*) Taylor and Williams, despite their commitment to historical philosophising, seem much less sensitive to the actual weight of that tradition. Or perhaps, sensing its hegemony, they react too hard against it.

Again a historical perspective is useful. It helps one to grasp just how deeply entrenched, within the naturalistic outlook that now dominates modern societies, is a certain conception of the good: the idea that the good can only be the impartially considered good of individuals. I think only one other overarching ethical idea remains as deeply entrenched: the idea of moral self-governance. The development of each of these ideas can be traced through the modern period – J. B. Schneewind's study, *The Invention of Autonomy*, does this for the latter up to Kant.

Self-governance does not start from any overall conception of the good; rather, it gives morality itself a central meaning in life and insists that its obligations or 'musts' are grasped directly by the deepest personal insight. It can be given religious, Kantian, or existentialist directions. It elevates such concepts as conscience, autonomy, authenticity, integrity – the lists suggest a rough historical order. The tension between these two ideas is a significant force in modern ethical thought (see 'Welfare and Self-Governance').

Utilitarianism is at its strongest as one powerful expression of the individualist and impartial conception of the good. Individualism and impartiality are also pillars of modern, particularly of democratic, politics. There are live alternatives: lowbrow Hobbesian ones within the naturalistic outlook, and highbrow eudaimonist or idealist ones which may conflict with that outlook. But I do not think either of these can capture the *fundamental* status, to us, of the idea that everyone is the subject of absolute ethical concern, no-one more so than anyone else.

Within the impartialist camp however there is the Kantian alternative to utilitarianism. Its appeal to the equal worth, dignity, or respect due to all human beings is obviously in tune with the democratic ethos – more so in some ways than utilitarianism. Self-governance, not a theory of the good, is the cornerstone of the Kantian approach. It attempts to get equal respect and morality itself from the very idea of our transcendental, absolute ability to give ourselves the law. Alas, I agree with Hegel's well-known assessment, that Kant deserves the highest honour for his insight into autonomy, but that the project of deriving morality from the idea of ourselves as autonomous, via the 'Categorical Imperative,' is a complete failure. And despite the importance of autonomy (both epistemological and moral) Kant's absolutism about it also seems to me to be misguided. Relations of equal respect are indispensable to democratic civility. Beyond that 'equal worth' is our modern noble myth, like the mediaeval noble myth of chivalry. Ethics needs to be much more down-to-earth, much more realistic, about people; it must go beyond noble myths, not feed them.

Impartiality itself, however, depends on no myth and remains a cornerstone of modern ethics. But utilitarianism is only one version of it. There are conceptions of impartiality other than sumtotal, and conceptions of individual good other than hedonistic. Our increasing awareness of the space of alternatives in both respects seems to me to be one of the most interesting and important developments in recent moral philosophy. I come back to it below.

One other great issue for contemporary ethics is the status of morality. Formally egoistic and formally impartialist moral philosophies have both tried in their various forms to find a way of subordinating morality to their standpoint. But what if it cannot be subordinated? Should it be superseded, or acknowledged as an independent normative source? Tim Scanlon's *What We Owe to Each Other* is an important version of the latter approach. He seeks to displace the dominant role of utilitarianism within the modern perspective by appeal to a conception of morality which can suit that individualist and impartialist perspective equally well, without invoking a welfarist theory of the good.

Scanlon shows how powerful is the idea of testing moral principles against what a concretely placed individual could reasonably reject. I think it is powerful because the notion of reasonable agreement – reasonableness being understood as fair-minded realism in considering what people can be expected to take on – is indeed one key factor in the way moral common sense in a rational community is formed and reformed. Scanlon sees it as a version of contractualism. To me, though, it seems clearer to think in terms of dialogical convergence rather than contract: Hegel rather than Rousseau, so to speak. I would also say that utilitarianism and Scanlonian contractualism are each unrealistic about a different thing. Against utilitarianism, it is unrealistic to suppose that one can *derive* morality, old or new, from a theory of the good.

Against Scanlon's contractualism, it is unrealistic to ignore the significance for morality of a conception of the good. Welfarist ideas about the good continuously shape and revise our moral notions; it is a question of not over-simplifying the normative connections between them.

The terrain that emerges from this discussion, in my view, was roughly mapped by Sidgwick in *The Methods of Ethics*. To improve his map we must broaden 'egoism' to include all the agent-relative or partial preoccupations and commitments that largely make up personal life, 'utilitarianism' to the whole class of individualist and impartialist theories of the good, and 'intuitionism' to the class of views that start within morality, including in this Scanlonian contractualism (because – an observation not a criticism – 'reasonableness' is an ineliminably moral notion). Sidgwick sees a tension between the first two. Williams was also concerned with this tension – but wrongly in my view he gave up on the theory of the good. Sidgwick did not see that there is a tension between *both* of these standpoints and morality; he was insufficiently sensitive to the way that utilitarianism (or any impartialist standpoint on the good) and common sense morality tend to pull apart. Williams does see that, but – to schematise drastically – gives up on morality too. (As with Nietzsche, of course, although much more plausibly in Williams' case, one can alternatively argue that he presents a new *conception* of morality.)

My own view is that one cannot and should not give up any of these three normative sources, not even if they cannot be fully reconciled. The question is how they flow into conclusions about one's reasons for action. I hope to address this question, at least to some extent, in the book about reasons I mentioned earlier.

### How do studies within scientific disciplines contribute to the development of normative ethics?

Natural science obviously contributes in a major way by posing wholly new moral problems that arise from the technological advances it has brought. But I take it the question is whether scientific disciplines can contribute relevant information, new ideas, or new methods to ethics.

At the broadest level the natural sciences contribute by reinforcing our naturalistic self-conception. In more specific ways they seem to me as yet to have few really solid implications. Biology and psychology may in due course tell us something about empirical limits and variations of moral competence, or of moral responsibility; but they haven't done that in any fully theorised way yet and may never do. They haven't, that is, shown us anything that history and intelligent common sense, as against wishful self-images, does not show us (for example about how easy it is to manipulate people into doing evil), though they may have confirmed it. Similarly, in my judgement they haven't, so far at least, shown us anything really unexpected about human motives, or imposed new conceptual frameworks on common-sense moral psychology.

I would like to think that anthropology and history give us interesting comparative material, and of course this is to some considerable extent true if one takes, again, a wide enough perspective. Unfortunately however any philosopher who has specific questions to ask about particular moral concepts, for example about the development or ubiquity of the notion of moral responsibility, as against other notions of agency, finds it is very hard to get sufficiently exact information from them. That is not a complaint; anthropological and historical texts, especially contemporary texts, have not usually been written with such philosopher's questions in mind. It would take a collaborative project to get the right focus.

Formal methods in decision theory have been helpful in providing clear axiomatic treatments of some basic ideas about individual and general good. I have found it very helpful, for example, to read discussions of the 'sure-thing principle.' The debate between Allais and defenders of expected utility theory like Savage seems to me a model of how experimental results can bring epistemic pressure to bear on strongly held normative convictions. (Though assuredly they cannot conclusively refute them.) However, the usefulness of formal methods, whether at the normative or the meta-normative level, is easily exaggerated: we should not trim the questions we ask to these ways of answering them, or treat the ones that can be so trimmed as more important for that reason.

**What do you consider the most neglected topics and/or contributions in normative ethics? And what are the most important problems in normative ethics and what are the prospects for progress?**

The world of philosophy has become so big that someone seems to be writing about just about everything imaginable. One might almost ask what topics are not being neglected enough! Of course

if some important topics are presently neglected I may well be one
of the people neglecting them.

At any rate what one considers neglected is influenced by what
one thinks important. So I am putting questions 4 and 5 together,
and will say something about the following list of topics that seem
to me important, starting with the easiest ones for an academic
philosopher to say something about, and ascending in order of
difficulty, neglect, and arguably ethical importance:

(i) Post-utilitarianism

(ii) The significance and authority of morality

(iii) Ethical criticism of our values and ideals.

**(i) Post-utilitarianism.** I have already said that an individualist and impartialist conception of the good seems to have become
a fixture in our ethical world; rightly in my view. However it is no
longer possible for traditional utilitarianism to dominate it as its
sole expression.

We see much more clearly how the good so conceived can have
any one of many plausible distributive structures. We also see
how hard it is to remedy the counter-intuitive aspects of any of
these structures. A notorious weakness of traditional utilitarianism, considered as a theory of the good, is its acceptance of un-
limited trade-offs. Few people would regard a small increase in
sum-total well-being achieved at the cost of serious suffering for a
small number to be an *improvement*. Versions of prioritarianism
can mitigate this problem but not eliminate it (in a way that seems
plausible overall). Alternative approaches, involving thresholds or
discontinuities in the function from the good of individuals to the
general good tend to produce their own counter-intuitive results.
(I have put forward one of these – see 'Value and Distribution' in
*Ethical Explorations.*) Then there is the question of how to relate
questions about the distributive structure of the good to questions about moral concepts such as fairness and desert. This is an
area in which much thinking remains to be done; there should be
no a priori assumption that the results will be determinate and
smooth.

A similar enlargement has occurred in relation to questions
about the constituents of human good. Mill's method, of defining
these in terms of what is ultimately desirable to human beings
and making the test of what is desirable what is actually desired
in reflective practice, is perfectly correct. In my view he was also

right to subordinate ideals of excellence to well-being (as I think he did – even though he attributed so much importance to ideals and wrote so eloquently about them). Ideals that diminish rather than deepening well-being should fall. But Mill was wrong to think that by his test happiness is the only thing desirable. People resiliently desire things other than happiness, such as knowledge of their situation, the good of people and causes with which they identify, freedom to live their life in their own way. They are willing to give up some happiness to achieve these. To explore these ends with appropriate imagination and depth seems to me an important thing some moral philosophers might now do. James Griffin provided an important lead in *Well-Being*.

**(ii) The significance and authority of morality**. In the $20^{th}$ century quite a few intelligent people came to think of morality as a form of self-repression; the 'peculiar institution' in Bernard Williams' words (1985). Obviously an important source was the stream of sexual liberation from Freud to the 60s; but there are deeper philosophical criticisms. They come from two sides: welfarist and Nietzschean. What good, from the welfarist standpoint, do the moral notions do – especially when we see that they are not derivable from it? May they not be positively dysfunctional? Nietzsche's critique of those notions – conscience, guilt, remorse, moral responsibility, blame and punishment – is of course quite different. It is grounded in an overwhelming, philosophically and historically rooted, liberationist and anti-welfarist story. Yet the two sides easily blend.

What should one make of this? First of all, it could be said that this suspicion is itself by now a somewhat antique posture. Do current forms of social conformism still centre on moral self-repression, as Mill thought the conformism of his fellow Victorians did? Critics of morality like Williams – when considered as critics of our culture – are in this respect generals fighting a previous war. Still, that is so far a cultural not a philosophical response. Philosophy has to respond to the deeper philosophical criticisms. It has to show that conscience and responsibility, blame and reconciliation are not repressive instruments but essential aspects of living freely together.

It is a large task. On the one hand it involves reasserting and explaining the role played by will and practical reason in motivating and justifying moral action, against those who think that will and practical reason are fictions. On the other hand, even more importantly, it calls for a realistic positive account of what moral valuations are and what role they play in our life. It seems to me particularly unhelpful merely to assert the unique and indefinable nature of moral obligation, as some intuitionists are wont to do. That only adds to the air of mysterious perversity.

Many moral philosophers have given illuminating down-to-earth accounts of the moral sentiments, from Adam Smith in *The Theory of the Moral Sentiments* to Alan Gibbard in *Wise Choices, Apt Feelings*. My own view differs somewhat from these illustrious forebears. I hold that the sentiments distinctive of morality are not at bottom those of resentment, indignation or fear; instead they are characterised by reciprocal dispositions of withdrawal of recognition and reconciliation (at-one-ment). Recognition, of oneself by oneself as well as by others, gives dignity, value and structure to a life. Hence its withdrawal can be frighteningly disorienting. The sentiments involved in morality are extremely powerful – and they are all too easily coldly manipulated and suborned, or wholly overpowered by resentment, rage, moral panic. It is however a grave error to make these emotions the essence of morality. If you confuse the patient with the patient's disease, it's not surprising that you try to kill the patient. (See Part III of *Ethical Explorations*, also 'Blame, respect and recognition' and 'Internal reasons and the scope of blame.')

**(iii) Ethical criticism of our values and our ideals.** Finally to one of the most important roles of normative ethics – ethical criticism of our values and our ideals, substantive discussion of what makes life truly worth living.

I think that philosophers, certainly including me, nowadays feel constrained about engaging in such criticism. Apart from anything else, it is very difficult to do, and the abilities it calls on do not necessarily correlate with those fostered by analytic philosophy. But I doubt whether that is all there is to it. It is in fact an interesting question why there should now be this constraint. The popularity of ethically neutralist stances in political philosophy is a telling symptom of it.

Ethical criticism and revaluation was one important thing philosophers did in the ancient world, from Socrates on. In the Christian centuries it became a task of priests and theologians; from the later enlightenment onwards it once again became a philosopher's concern – in the $19^{th}$ century no important moral philosopher failed to address it, and most of our still active ideals date back to that time. So where are we now? It is not that people in 'Western' liberal democracies show a lack of moral concern about urgent moral

issues such as poverty, oppression, global warming. I am raising a
different question. Are we living off certain ethical ideals without
really being willing to defend or revise them, or even scrutinise
them? And if so, why should this be? Does it matter? Charles
Taylor's work, especially in *Sources of the Self*, raises the question
with admirable depth, but still stands in splendid isolation, and
presents only one possible approach. It would be desirable indeed
to have much more discussion.

## Bibliography

Gibbard, Alan, *Wise Choices, Apt Feelings* (Oxford: Oxford University Press, 1990).

Griffin, James, *Well-Being* (Oxford: Clarendon Press, 1986).

Hegel, G. W. F., *Elements of the Philosophy of Right* (Cambridge; Cambridge University Press, 1991).

Sidgwick, Henry, *The Methods of Ethics* (London: Macmillan, 1907).

Scanlon, T. M., "Contractualism and utilitarianism", in A. Sen and B. Williams (eds.), *Utilitarianism and Beyond.* Cambridge: Cambridge University Press, 1982).

Scanlon, T. M., *What We Owe to Each Other.* Cambridge, Mass: Harvard University Press, 1998).

Schneewind, J. B., *The Invention of Autonomy* (Cambridge: Cambridge University Press, 1998).

Skorupski, John, *Symbol and Theory* (Cambridge: Cambridge University Press, 1975).

Skorupski, John, *Ethical Explorations* (Oxford: Oxford University Press, 1999).

Skorupski, John, 'Blame, respect and recognition', *Utilitas* 17, 2005.

Skorupski, John, "Propositions about reasons" *The European Journal of Philosophy*, 14, 2006.

Skorupski, John, *Why Read Mill Today?* (London: Routledge, 2006).

Skorupski, John, "Welfare and Self-Governance," *Ethical Theory and Moral Practice*, 9, 2006.

Skorupski, John, "Internal reasons and the scope of blame," in Alan Thomas (ed.), *Bernard Williams* (Cambridge: Cambridge University Press, 2007).

Smith, Adam, *The Theory of the Moral Sentiments*, D.D. Raphael and A.L. Macfie (eds.) (Indianapolis: Liberty Fund, 1982).

Taylor, Charles, *Sources of the Self* (Cambridge: Cambridge University Press, 1989).

Williams, Bernard, *Ethics and the Limits of Philosophy* (London: Fontana Paperbacks, 1985).

# 14

# Michael A. Slote

UST Professor of Ethics

University of Miami, USA

## Some Thoughts for the Future

When I first started thinking about philosophy, people interested in ethics did meta-ethics, and normative ethics was hardly visible as a discipline. But John Rawls changed all that, and because I was a graduate student at Harvard, I had the pleasant and useful opportunity to fall under his influence—even though I didn't actually begin to work in normative ethics till many years later.

In this short space I don't propose to offer any sort of account of what normative ethics is. I don't think I can define the notion, and I think some of the other contributors to this volume will have more considered views on that issue; views that I am confident will be expressed in some of the other essays here. However, I am fairly sure that most of the other contributors will think that normative ethics centrally involves ideas or theories about what is (morally) right or wrong. I believe they would also agree that questions about the good life and/or about what kinds of things/entities are good for their own sake are also part of normative ethics. But the kinds of answers that are currently given to questions of these kinds are enormously various. Normative ethics is a very rich and/or complex field at the moment and shows no signs of becoming less so in the future.

Also, there have been new developments over the past couple of decades, so that there is probably a greater variety of basic approaches to moral questions than there was not very long ago. This might be interpreted as indicating a lack of progress in the field, and I want to return to that question a bit later. But the greater variety I have in mind includes the recent revival of Aristotelian and other forms of virtue ethics and the development of what is called the "ethics of care" or "care ethics." I myself work or have worked in both virtue ethics and care ethics, and I would like to say something about how and why these new approaches emerged, before I turn to issues about the overall state of normative ethics and its prospects for the future.

The supposed failings of Kantian and utilitarian moral theory are often mentioned as the principal reason for favoring virtue ethics, which sees issues of right and wrong action as primarily depending on questions about the character or motivation of agents rather than on universal rules governing action or on the consequences of given (classes of) actions. And virtue ethicists have also claimed that (following Aristotle or Hume) they can offer a better picture of an individual's relations with his friends, family, or polity than anything to be found within the utilitarian or Kantian tradition. Interestingly enough, even though care ethics is most frequently not regarded as a form of virtue ethics, it makes rather similar criticisms of Kant and utilitarianism.

Now I don't think I have the space to discuss how valid those criticisms are and to consider, in particular, how Kantians and utilitarians would respond to them. (Not to mention the *advantages* they would respectively claim over virtue ethics and care ethics.) But I do want to say something about the origin and character of the ethics of care, since that is the kind of normative ethics I favor and since I think it is in fact unlikely that any of the other contributors to this volume will focus their attention on care ethics. I am also a believer in virtue ethics, but I don't think it is obvious that care ethics has to be inconsistent with virtue ethics.

Care ethics began in 1982 with the publication of Carol Gilligan's seminal work *In a Different Voice: Psychological Theory and Women's Development*. Gilligan argued that men and women tend to approach moral problems differently: men typically focus on issues of justice, rights, and autonomy and think in terms of the rational (and just) application of rules, laws, or principles to problem cases. Women more frequently (than men) deal with moral issues through the prism of their concern for others. Laws and rules are often less important for them than the connection they feel and have for other individuals as individuals.

Now the first thing that needs to be said about Gilligan's view is that it has, from the start, been very controversial. Anecdotally, or in terms of our own experience, there seems to be *something* to the distinction Gilligan made, but there is a vast literature questioning the empirical studies Gilligan and others have used to support the claim that men and women tend to think differently about moral issues. There is also an equally vast literature supporting Gilligan's original claims or at least qualified versions of them.

At the time Gilligan wrote her book, Freud, Piaget, and Lawrence Kohlberg had all fairly recently been saying that male thinking about morality is typically superior to that of women. But the studies her Harvard colleague Kohlberg had done on moral development had in fact been based solely on a sample of males. (This is also true of Piaget, who exercised an enormous influence on Kohlberg.) Kohlberg's theory said there were six stages of development, and when he applied it to females, it turned out that females typically advanced less far through those stages than men tend to do. This entailed or implied that men are on average morally superior to women, but Gilligan pointed out that, given his methodology, Kohlberg could or should only make the less strong claim that women's moral development *differs* from that of men. The importance she placed on that qualification or criticism is visible, moreover, in the title she chose for her book. Just because women don't on the whole follow the same stages of moral development that men do doesn't mean that they are morally inferior: they're just *different.*

And if women (to some extent) differ morally from men, it might well be worthwhile to understand, describe, and analyze what a distinctively female morality involves. The idea of an ethics of care is mentioned only briefly by Gilligan, but others have subsequently written a great deal about how an ethics of care should be articulated, and care ethics is now one of the major approaches to normative ethics. But the whole idea of such an ethics raises some enormous questions. How much territory, for example, is care ethics supposed to cover? Gilligan spoke initially of a contrast between thinking in terms of justice and thinking in terms of caring or connection, but if men think in terms of justice (and autonomy and rights) does that mean that an ethics of care has to leave that topic (those topics) aside?

This problem has been addressed by many care ethicists, and a variety of answers has been offered. Some have held that the ethics of care should confine itself to the non-political sphere and rely on standard "male" treatments and thinking for answers to questions about justice and rights (including rights of autonomy). On this view, roughly, care ethics and traditional male ethical thinking complement one another and can somehow be harmonized or integrated—even if women tend to think more or better about certain sorts of moral issues and men more or better about others. Other care ethicists have argued that women have or can develop their own distinctive way of understanding autonomy and even justice, and these ways, it has sometimes been said, are superior to what standard "male" moral theorizing has had to say about those topics. This is the way I *myself* have gone, though with one major proviso.

There is evidence that many women think in terms of traditional justice and autonomy, even if very, very few males approach morality in terms of caring (as Gilligan described it). So I don't think the difference between care ethics and traditional or standard thinking about justice, autonomy, and rights correlates very well with gender. Rather than think of care ethics as the ethics of women or as a female ethics, therefore, I think it best simply to see it as an alternative to certain forms of traditional thinking about justice, etc., most notably to the Kantian/liberal approach to these topics that is currently so dominant. (Gilligan herself is inclined to think of things in this way.) And in my present work I have been arguing that the liberal and care-ethical views of autonomy and justice are mutually inconsistent regarding whole classes of cases. Traditional and care-ethical thinking about these topics are, therefore, not harmonizable within some larger systematic morality, and if we are looking for a satisfying overall morality, we have to choose *between* liberalism, and the way it understands justice, rights, and autonomy, and what care ethics has to say about the whole range of moral topics. Let me explain why.

In what Joel Feinberg has usefully called "Skokie-type cases," liberalism argues that neo-Nazis' or others' rights of free expression trump the mental anguish and quite possibly even trauma that such speech would likely cause, say, in Holocaust survivors living near to where the neo-Nazis are proposing to demonstrate. Most feminists sharply disagree with this, and I think anyone holding an ethics of care would want to interpret autonomy in such a way that, in Skokie-type and other cases, it doesn't entail the broader or strong notion of our (rights of) autonomy that liberals subscribe to. Liberalism, in other words, defends the right to give expression to hate speech in Skokie-type cases, but an ethics of care would say that the fact that such hate speech is likely to cause anguish and trauma morally trumps the considerations that favor allowing such speech. We really do, therefore, have to choose between what liberalism tells us and what care ethics tells us, and the task, then, for the care ethicist, but also for the liberal, is to give some good or objective reason or argument for preferring their particular view of Skokie-type cases. If either side can do this, that will favor it as an comprehensive approach over the other, and I feel, therefore, that the issue of hate speech is both crucial to the prospects of care ethics and central to current normative ethical theory in a way that is not generally recognized.

The importance of these issues connects with the importance of Gilligan's findings and of the whole idea/challenge of an ethics of care, but, interestingly and (I would say) depressingly enough, contemporary liberals and Kantians (and, for that matter, utilitarians and Aristotelian virtue ethicists) don't take this whole nexus of thought and theory very seriously. Going by what you read in work done in those traditions, it is practically as if Gilligan had never written. Within "mainstream" normative ethics that contribution is very largely, and, I would say, very unfairly neglected in favor of business as usual.

I am not sure I know why this should be the case. Kuhn points out that adherents of an old paradigm invariably reject any new paradigm when it comes into view, even if the old paradigm is in crisis, is faced with troubling problems it has no idea how to solve or even deal with. And the current scene in normative ethics seems in any event different from what Kuhn describes as happening in science for a number of related reasons. Ethics isn't science, for one thing, and for another it is hardly clear that the ethics of care is going to eventually drive Kantian liberalism from the debating chamber. Because of the stress philosophy itself places on reason and rationality, rationalistic views have a certain natural appeal within the field of normative ethics, and Kantian liberalism is rationalism par excellence, while care ethics is deliberately and explicitly opposed to rationalism and sees itself as part of (and fulfilling?) the sentimentalist tradition of Hume and Hutcheson. I don't see the ethics of care taking over the field of normative ethics—but, then again, I don't see any one approach taking over the field. This is philosophy and philosophy doesn't develop, or progress, in the way science does, though there is no time to say anything more about this here.

In further disanalogy with what Kuhn says about periods of scientific revolution, the Kantian liberalism that currently predominates in normative ethics doesn't seem to be facing a crisis—though that appearance may be deceiving. What we are seeing now, within ethical rationalism, is intensification, a heating up, if you will, of its commitment to a rationalistic understanding of ethical phenomena. When Thomas Nagel, in *The Possibility of Altruism*, first suggested that reason, rather than desire, can often motivate what we do, his arguments were intuitively forceful and constituted a real challenge to the widely accepted Humean desire-focused view of the explanation of action. But nowadays the rationalism has become much more extreme, and we have T.M. Scanlon, for example, arguing that the most basic ordinary human actions involve our being responsive to what appear to us to be reasons to do one thing rather than another. What is ordinarily thought of as (just) desire has become hyper-intellectualized into something entirely within the rational sphere. Scanlon acknowledges a major debt to Nagel, but what he and other rationalists working on issues of moral psychology have done is to remove the account of human action and motivation from anything related to empirical or scientific psychology and to treat the basic elements of action as entirely understandable in terms of distinctions of reason or rationality.

One also sees this tendency in rationalist accounts of feeling or emotion. Kant thought we should act from reason rather than from emotions or feelings like compassion or love, and these latter were not themselves seen as purely or even partly rational—which was a major part of Kant's rationalistic objection to their influence upon action. But I recently heard a paper by David Velleman in which he argued that far from being separate and distinct from the rational side of our natures, love is best understood (roughly) as the expression and embodiment of the value we place on the moral worth and dignity of other individuals as rational beings. This is another good example of the recent tendency, on the part of mainstream ethical rationalists, to see every element of human psychology in constitutively rational terms, and that is why I say rationalism is becoming more extreme.

Is this a crisis? Well, the rationalist obviously doesn't think so. From the rationalist side, the appearance, I'll bet, is that rationalism is going from strength to strength as it shows its ability to understand or account for more and more areas of the mind. But this period of intensifying rationalism also corresponds to a period in which rationalism has been increasingly challenged and criticized for over-intellectualizing, over-rationalizing human life and activity. The most philosophically famous example of such criticism is Bernard Williams's "one thought too many" objection to Kant's advocacy of universal conscientiousness. But Gilligan and the ethics of care clearly represent another—and rather more systematic—challenge to rationalism, and is it possible that the recent extremism of rationalism represents a frenetic reaction to the challenges and criticisms? Is it possible that the rationalists under fire prefer to increase speed rather than change course or turn back? If this were so, the reaction would be understandable. It's the kind of thing human beings often do—for example, in the sphere of international politics. But that doesn't mean that, either in normative ethics or in politics, it is a good idea to react in this way.

Right now the rationalists are largely ignoring (or desultorily swatting at) those, like care ethicists, who largely question their tradition. But the fault seems hardly to be all on the one side. Ethicists of care have ignored some of the most important questions of normative ethics: for example, the problem of how to justify deontological restrictions on the pursuit of good results and moral problems concerning legislation and constitutional design. And these are questions that rationalist Kantians and liberals work hard on and have interesting ideas about. So what we seem to have now, in fact, is two traditions ignoring each other or each other's issues. Since these two traditions correspond (very roughly, since many prominent Kantians and liberals are women) to the distinction Gilligan made between male and female approaches to moral issues, one might say that these recent historical developments are further evidence for the truth of what Gilligan was saying. But to that extent they may also justify a certain pessimism about the kind of progress that can or cannot be made in normative ethics. They certainly can make one wonder whether there is such a thing as a universally valid human morality or at least whether we are capable of knowing what it is. But in any event these are questions that ought to bother both liberal Kantian rationalists and care ethicists, and the former, at least, don't seem to *be* bothered.

This can, as I said, make one feel somewhat pessimistic. But the facts on the ground can also be seen as a call to action. Those, like myself, who think what Gilligan has said is important may want and try to see whether an ethics of care can deal with some of the issues that till now, or recently, have been the exclusive purview of Kantian liberals and utilitarians. (Aristotelian virtue ethicists have also tended to ignore the theoretical problems I have mentioned.) And by way of drawing this essay to a close, let me focus briefly on how I think the ethics of care might attempt to deal with the central normative issue of deontology.

Deontology is, roughly, the view that there is something wrong with certain kinds of actions—killing the innocent, theft, deceit—even apart from their (usual) bad consequences, and most normative ethical theories accept and/or seek to vindicate deontology because of the deep-seated intuitions most of us have that deontology has to be true. Rationalists either claim that deontology is intuitively valid and needs no (further) vindication or give elaborate philosophical arguments in defense of our deontological intuitions about particular cases. No appeal is (usually—Rawls is a notable exception) made to psychology or other sciences either in regard to this problem or in regard to other ethical issues.

But born out of the field of educational psychology as it is, the ethics of care has tended to see psychology as relevant to its claims. However, care ethicists have not yet seen the potential relevance of psychology to the question of deontology, and let me just say a bit here about how that might turn out to be possible. Care ethicists have said a great deal about the importance of caring, but less about how caring develops. But psychologists have done a great many studies of moral development in recent decades, and one thing that emerges from that literature is that genuine altruism—which is another name for caring—seems to depend and thrive on the development of empathy (roughly, feeling what some other person or, possibly, group feels). Psychologists have traced the course of empathic development within children and adolescents (idealistic empathic identification with disadvantaged or oppressed groups emerges during the teenage years), and the majority view, resulting from many empirical studies and much discussion, seems to be that empathy is a major force behind our concern for, or caring about, others.

The psychologists also note our tendency to feel more empathy and concern for our intimates than for total strangers and for those whose distress or pain we witness rather than for people we merely know *about*. These distinctions in fact correspond to moral distinctions we intuitively want to make. We think our obligations to intimates are greater than to strangers, and likewise (*pace* Peter Singer) someone who fails to respond to the pain or danger of a child drowning right in front of him or her seems to be acting worse than someone who fails to give money to Oxfam that would save some child in a distant part of the world. If the deontological distinction between killing and letting die, or, more generally, the distinction between doing and allowing is one that empathy is sensitive to, that might at least suggest a way in which an ethics of care could defend our deontological intuitions. And I think empathy is sensitive to that distinction in something like the way it is sensitive to what it perceives rather merely knows about. Our empathy tends to be more aroused and we tend to care more about a pain or suffering that we see, as opposed to pain or suffering we merely know about (great literature and vivid descriptions can blur this distinction). But we similarly react more strongly to pain or suffering we know we might inflict or cause than to suffering we might merely allow to happen. We *flinch* from the possibility of killing someone much more than from allowing someone to die, and this emotional reaction seems a form of (heightened) empathy. So perhaps that is why we morally care more about not killing than about not letting die and similarly for others of our deontological intuitions and commitments. This doesn't yet amount to an argument that explicitly justifies deontology, but it makes a strong gesture in that direction, and the book I have just finished, *The Ethics of Care and Empathy*, seeks to make the justification or argument stick.

For the moment, though, this example shows how care ethics might seek to answer some difficult normative ethical questions making use of the kind of empirical studies and claims that rationalism by and large avoids. This isn't *necessarily* an advantage for care ethics, since much of our ethical thought seems to be a priori valid (the analogy between ethics and mathematics goes back at least as far as Plato). But a priori vindications of deontology have, historically speaking, been notoriously unsuccessful. So both rationalist liberal Kantianism and sentimentalist care ethics have work to do. However, the normative ethical climate or situation might be somewhat better or healthier, if both sides paid more attention to one another

## Bibliography

Carol Gilligan, *In a Different Voice*, Harvard, 1982.

Rosalind Hursthouse, *On Virtue Ethics*, Oxford, 1999.

Thomas Nagel, *The Possibility of Altruism*, Oxford, 1970.

John Rawls, *A Theory of Justice*, Harvard, 1971.

Michael Slote, *The Ethics of Care and Empathy*, Routledge, 2007.

T. M. Scanlon, *What We Owe to Each Other*, Harvard, 1998.

David Velleman, "Love as a Moral Emotion," *Ethics* 109, 1999, pp. 338–74.

Bernard Williams, "Persons, Character and Morality" in his *Moral Luck*, Cambridge, 1981.

# 15

# Wayne Sumner

University Professor

University of Toronto, Canada

**Why were you initially drawn to normative ethics?**

It was not always this way. I am a survivor of the era in which dealing with normative issues was considered beneath the dignity of serious philosophers, an appropriate occupation for journalists or preachers, perhaps, but not for deeper intellects. A friend of mine once observed that the philosophical styles we take most seriously are those that were in vogue when we were in graduate school. Well, I was in graduate school at Princeton in the early 1960's, when ethics was identified with metaethics and when the reigning theories were nonnaturalism and noncognitivism rather than consequentialism and deontology. For a while that was the game I too played, up to and including the writing of a metametaethical thesis on the relationship between normative ethics and metaethics. Soon after leaving Princeton and coming to Toronto, things changed. But I am getting ahead of myself.

When I was an undergraduate I had only the vaguest notion of what I wanted to do with my life, but I knew that, whatever it was, I wanted it to matter, to make some difference to the world. Unluckily for me, given that ambition, it soon became clear that the only thing I was really very good at was philosophy–on the surface of it, the most impractical of all disciplines. Once I drifted into philosophy, however, I very quickly migrated to ethics, due in part to its positive attraction (since it seemed the most practical part of philosophy) and in part to my manifest incompetence at metaphysics, epistemology, and any other heavy-duty, hard-core area within the discipline. Ethics, I naively thought, has to do with how we should live our lives, how we should treat others, how institutions should be designed and policies justified, and other such weighty matters. My thought was naive because, as was underlined once I arrived at Princeton, this was the age of metaethics and philosophers were not addressing any of these questions. I have already confessed to conforming with the dominant culture while in graduate school, but perhaps now it is clearer why my career as a metaethicist was doomed to failure from the outset.

Two things happened in the latter 1960's that changed everything for me. One was my discovery – or rediscovery – of utilitarianism. I had read Mill as an undergraduate, though not Bentham or Sidgwick, and Moore as a graduate student, though the emphasis was inevitably on the naturalistic fallacy and other aspects of Moore's nonnaturalism rather than his consequentialism and pluralist axiology. But I knew little of the further twentieth-century debates about utilitarianism, which somehow persisted despite the hegemony of metaethics. Then in 1965 David Lyons published *Forms and Limits of Utilitarianism*, by far the most thorough and systematic treatment to that time of the act v. rule utilitarianism debate. Lyons's book was followed two years later by D.H. Hodgson's *Consequences of Utilitarianism*, and suddenly it was fashionable once again to talk about normative theories. Not only that, but I quickly decided that Lyons and Hodgson were talking about *my* normative theory. Once I began thinking about it, I found that the pragmatism and empiricism of the utilitarian tradition perfectly suited my cast of mind. One day I woke up and decided – or discovered – that I was a utilitarian.

The other formative influence at the time was beyond the confines of academic philosophy. As everyone knows, or remembers, the latter part of the 1960's was a period of rapid social change, at least in western Europe and North America. The war in Vietnam was approaching its peak intensity, as was the antiwar movement, and the advent of feminism was placing issues of gender, sexuality, and reproduction at the centre of attention. These were all questions which engaged me as a citizen and activist, but they were disconnected from my intellectual life as a philosopher. That disconnect quickly became intolerable, especially because I had been attracted into ethics in the first place by its promise of having something to say about these very questions. So now I had a theory and I had some problems; all that remained was to bring them together.

By the end of the 1960's I had abandoned metaethics, never to return, and had begun working on the application of utilitarianism to the problem of abortion. At first I thought this would be a relatively simple matter, requiring no more than a journal article or two. More than a decade later I finally managed to work out my own views, more or less to my satisfaction, in *Abortion and Moral Theory* (1981). However, if I thought that matters would end there then I was once again being naive. In that book I adumbrated what I regarded as two key ingredients of a viable form of utilitarianism: a theory of welfare and an account of how a consequentialist moral structure might accommodate rights. Since both of these ingredients required much further work, I decided that my next project would be a utilitarian theory of rights. That project occupied me for the balance of the 1980's, by which time it had become too unwieldy to fit into one book. The theory therefore appeared in two installments, as *The Moral Foundation of Rights* (1987) and as *Welfare, Happiness, and Ethics* (1996).

Just to show that nothing much has changed for me since the 1960's, my most recent project, which emanated in *The Hateful and the Obscene* (2004), once again undertook a consequentialist treatment of a cluster of practical issues, this time on freedom of expression. Actually, though, I guess one thing has changed, at least a little. In my earlier years I was happy to label myself a utilitarian, whereas now I tend to identify myself more cautiously as a consequentialist. Utilitarianism, as everyone knows, is one variety – not the only one – of consequentialism. So I now tend to affiliate more with the genus than with the particular species. When I teach utilitarianism in my graduate seminars I decompose it into three essential ingredients: consequentialism, welfarism, and aggregation or sum-ranking. I tell my students that my confidence level in the first is 100%, in the second 90% (since I can at least appreciate the attraction of more pluralist axiologies), and in the third maybe 60% or 70% (since I have a stronger appreciation for the attractions of distributive principles). So that's a shift from the 1960's though, all things considered, not a very big one.

## What example(s) from your work (or the work of others) illustrates the role that normative ethics ought to play in moral philosophy?

My answer to this question is short and straightforward: normative ethics makes moral philosophy, and philosophy in general, relevant. Consider the basic questions of metaethics, concerning the cognitive status of moral judgements, or the supervenience of normative properties, or verification procedures for moral principles, or whatever. It would be a mistake to suggest that these issues are the exclusive concern of philosophers, as we are reminded every-time social scientists repeat the mantra that values are inevitably subjective or students in our introductory classes recite their ar-guments for cultural relativism. But they are some considerable distance from the ethical concerns most on the minds of ordinary folk, which have to do with their responsibilities to their children, or the legal recognition of same-sex marriage, or what we should all be doing to mitigate the worst effects of climate change. Most people's ethical concerns are normative, most of the time, and they expect philosophers to have something to say about these concerns. Indeed, the popular image of philosophy is that it is the one discipline willing to grapple with the Big Questions in life: What is the meaning or purpose of our existence? How should we live our lives? What do we owe to those closest to us, or to distant strangers, or to our descendants? These are all normative questions.

To the extent that philosophy has nothing to say about these questions then it disappoints public expectations of it. More im-portantly, perhaps, it may also disappoint political expectations of it. Real philosophy, as opposed to the stuff commonly found on the self-help shelves of bookstores, is taught primarily in universi-ties. Many of those universities are publicly funded, in which case hardworking people are being taxed to support this intellectual endeavour. How can this kind of coerced subsidy of our profes-sional lifestyle be justified? It may well be true that all philosoph-ical knowledge—assuming there is such a thing—is intrinsically valuable, so that there is some payoff to supporting philosophical inquiry into the nature of causation or the relationship between form and matter in Aristotle. But is this intrinsic value sufficient to warrant paying large numbers of metaphysicians fancy salaries to work nine months of the year? Shouldn't the government, and the public who finance it, have a right to expect somewhat more prac-tical returns from their investments? They can easily find these returns in other areas of the university—in medicine, engineering, law, the social sciences, and so on. But where are they to find them in philosophy, if not in the contribution the discipline can make to the great normative issues of public life?

If this is the best, or only, way of showing the practical relevance of philosophy, and if this relevance is a sine qua non for the contin-ued public support of the discipline, then three conclusions read-ily follow: philosophy needs ethics, ethics needs normative ethics, and normative ethics needs practical ethics. This last step may surprise, since practical (or applied) ethics is generally held in low regard within the discipline. And it must be conceded straight-away that a lot of the work in this field deserves no greater re-spect. On the other hand, if philosophy is to establish or maintain its relevance then it cannot limit normative inquiry to the theo-retical level. The average person on the street, who neither knows nor cares whether quasi-realism is the best account of the status of moral judgements, knows and cares as little whether consequen-tialism can accept an agent-relative axiology or whether Kant has a coherent argument to the humanity formula of the categorical imperative. The normative questions on most people's minds are not abstract and theoretical but concrete and practical. It is on these questions they look to philosophers for help. We would do well to provide it.

### How do studies within scientific disciplines contribute to the development of normative ethics?

Of the many ways in which the empirical sciences might be cru-cial to the success of the normative enterprise, I will focus here on just two. The first, and most obvious, is especially acute for those who, like me, employ a consequentialist methodology in dealing with practical normative questions. Consequentialism, which ba-sically enjoins us to prefer the option which will yield the best overall balance of benefits over costs, is notoriously information-dependent. Outcomes cannot be evaluated, and ranked, unless or until they can be projected with some degree of confidence. These projections require evidence – reliable evidence – and such evi-dence can normally be furnished only by the sciences.

I will illustrate this process by reference to my recent work on freedom of expression. In dealing with the challenges for free ex-pression provided by pornography and hate speech I applied a de-cision framework adapted from Mill's essay *On Liberty*. The first step in this framework is to determine whether the expression in question poses a significant risk of harm to others. If it does not then no interference in it by the state can be justified. But if it does then the second step is to determine whether state interfer-ence in the expression can be justified by means of a cost-benefit comparison in which the harms of the expression are balanced against the harms of regulating or prohibiting it. It will be obvi-ous how the first step requires reliable evidence of the effects of pornography and hate speech and how the second step requires

equally reliable evidence of the effects of state regulation or prohibition. Since the only sources of reliable evidence on the first
question are the social sciences, an entire chapter of *The Hateful and the Obscene* was devoted to surveying the social science
literature on the potential harms of pornography (principally to
women and children) and of hate speech (principally to visible
minorities). These harms then needed to be balanced against the
expected consequences of state interference in these forms of expression, taking into account both how effective that interference
was likely to be and what collateral impact it would be likely to
have on other forms of expression. In the course of conducting my
own inquiry – and my own cost-benefit balancing – I was critical
of the Canadian courts for their lenience in upholding state cen-
sorship on the basis of little evidence of the harms it was intended
to prevent or of its effectiveness in preventing them. Good public
policy, I argued, must rest on good evidence, and that evidence
can be found only in the sciences, not in armchair speculation or
common sense.

This kind of reliance of ethics (especially applied ethics) on science is not surprising. But there is another way in which science
is implicated in the normative enterprise, this time on a more theoretical level. Whatever other demands we might place on normative theories, they surely must be capable of guiding the decision-
making of moral agents. Since these agents are human, it seems
reasonable to expect a good theory to be based on empirically
sound assumptions about human nature. It would count, therefore, against a normative theory if its assumptions about agents
were unrealistic or lacking in empirical support. Consequentialist
theories are often criticized on this basis, especially for imposing
greater decison-making burdens on agents than they can realisti-
cally be expected to bear. Consequentialism takes an impersonal
perspective on moral decision-making by telling agents that costs
and benefits are to be compared without reference to whose they
are (each is to count for one, no one for more than one). When
I am faced with a moral choice, therefore, the implications for
myself, and for those nearest and dearest to me, must be given
no more weight than the effects on distant strangers. To many
critics this has seemed excessively demanding, or even inhuman,
in requiring us to ignore or suppress our natural tendency to develop emotional ties to particular others: family members, lovers,
friends, fellow citizens, etc. If this criticism is well founded then at
least the simplest forms of consequentialism are a bad fit with our

nature and must be either abandoned or modified (perhaps by incorporating nonconsequentialist elements in their decision rules). But whether the criticism is well founded depends on the best account of what can be expected of actual human agents in their moral decision-making. And that account will be supplied by the sciences.

We should expect much of the evidence on this issue to come from the social sciences, which are capable of studying human decision-making processes and providing evidence of cognitive limitations, tendencies toward irrationality or bias, responsiveness to situational cues, and the like. In addition, evolutionary biology might provide an explanation for our tendency to favour members of our own kingroup over strangers. But some fascinating evidence may also be available in the neurosciences. Neuroimaging techniques appear to show that two different areas of the cortex are engaged in moral decision-making, one tending toward impartial cost-benefit calculation and the other bringing into play emotional attachments to particular others.[1] These two kinds of processes will conflict in certain high-conflict scenarios where personal harm must be done to an assignable person in order to prevent harm to a greater number of others (e.g., where the only way to stop a runaway trolley from killing five people is to throw a fat man onto the track). A recent study has shown that research subjects with focal damage in the ventromedial prefrontal cortex, the area which controls emotional responses, were more likely to lack the inhibition against harming the one in order to save the many and were therefore more likely to engage in straightforward consequentialist reasoning (minimize total loss of life) in these high-conflict scenarios.[2] One conclusion that could be drawn from this study is that if only brain-damaged individuals behave as straightforward consequentialists then consequentialism must be a bad fit with normal human nature. Alternatively, one could conclude that if human brains have evolved so as to balance these two different (agent-relative and agent-neutral) factors then that is a fact that consequentialists must take into account in recommending a decision-making procedure. That conclusion need not drive consequentialists to abandon their theory as an account of

---

[1] See, for instance, J.D. Greene et al., "An fMRI Study of Emotional Engagement in Moral Judgment", *Science* 293 (2001): 2105-2108.

[2] Michael Koenigs et al., "Damage to the Prefrontal Cortex Increases Utilitarian Moral Judgements", *Nature*, online edition, doi:10.1038/nature05631 (21 March 2007).

moral truth, but it might require them to abandon a straightfor-ward decision procedure in favour of one which incorporates some agent-relative inhibitions on harming others, simply as a de facto recognition of the force of these inhibitions in our moral thinking.

## What do you consider the most neglected topics and/or contributions in normative ethics?

Earlier I reflected on how normative ethics used to be a neglected area within ethics, and on how applied ethics used to be a ne-glected area within normative ethics. Neither claim could credibly be made now. The past forty years or so has witnessed an exponen-tial growth of work both on normative theories and on normative problems. Ethics journals flourish, indeed multiply, as do special-ized journals in the major applied areas (especially bioethics and environmental ethics, but also including social, political, and eco-nomic issues). It has to be conceded that much of the work that appears in these venues, especially the specialist applied journals, remains deplorably mediocre, but some of it–the best of it–is done by the best ethical minds of our generation. What remains ne-glected nowadays is harder to say, especially since (as Louis Al-thusser used to insist) we can't see around our own corners. So I will limit myself to one complaint: what normative ethics tends to neglect is its own past.

Orthodoxy now has it that there are three main kinds of nor-mative theory: consequentialist, deontological, and virtue-centred. (Speaking personally, I think that any viable form of virtue-centred theory will reduce to one of the other two, or a combination of them, but let that pass.) I don't mean to claim that advocates of these three theories lack a sense of their own history. Conse-quentialists standardly trace their ancestry back to Bentham and Mill (and perhaps also to Hume), deontologists regularly invoke either Kant or their natural-law roots in Aquinas, and virtue the-orists cannot be accused of neglecting Aristotle (more's the pity). What I have in mind instead is the tendency among philosophers to think that the latest construction of the main issues concern-ing their favourite theory is either entirely novel or demonstrably superior to anything their predecessors may have come up with.

I will confine myself here to one instance of this sweeping gener-alization. In contemporary analytic ethics much of the conversa-tion between consequentialists and deontologists is conducted in the language of agent-neutral v. agent-relative reasons. I myself accept this as an illuminating way of thinking about the issues that divide these two quite different approaches to doing normative ethics. However, there seems little recognition of the fact that, though the terminology itself may date back only to the work of Thomas Nagel and Derek Parfit in the 1970s, the basic idea behind the distinction is to be found a century earlier in Sidgwick's *Methods of Ethics* and was well known in the early decades of the last century, for example in Broad's analysis of 'self-referential altruism'.[3] It may, of course, be true that the latest iteration of the idea has been worked up to a much higher level of sophistication than anything that went before, but how can we tell if we ignore what went before? In the sciences, where theory confirmation may be a more settled matter, there may be more justification for thinking that the best theory is always the latest one and therefore that there is little to be learned from looking at past efforts (though this surely does not apply to the human sciences). But philosophy, or at least ethics, has tended to cycle repeatedly through basically the same ideas over the past couple of centuries, so unless we think that we are just a whole lot smarter than our predecessors it might dawn on us that we have something to learn from the way they handled these ideas.

Normative ethics seems to operate with a rolling wall of about thirty or forty years. So currently anything back beyond 1970 or so is pretty well lost to view. The past comes back into view when it is about a century old and then whatever has survived tends to remain in our consciousness, at least as prehistory. So we have recently recovered Sidgwick and Moore is back in fashion, but only a few diehards (like my colleague Tom Hurka) take Ross or Ewing seriously. Likewise, in metaethics the work of Richard Hare, once so enormously influential, has fallen off the philosophical agenda, presumably to be recovered only upon the centenary in 2052 of the publication of *The Language of Morals*. Quite apart from the disrespect to our predecessors evinced by this historical myopia, it offends me because it is inefficient, requiring us always to invest additional labour in reinvention when we could instead appropriate and adapt. Furthermore, it smacks to me of a form of philosophical hubris which we have done nothing to earn. Are we really that much smarter than our philosophical ancestors (who

---

[3] C.D. Broad, "Certain Features in Moore's Ethical Doctrines", in Paul Arthur Schilpp (ed.), *The Philosophy of G.E. Moore* (La Salle, Illinois: Open Court, 1942), pp. 43-67.

doubtless had the same condescending attitude toward their ancestors)?

## What are the most important problems in normative ethics and what are the prospects for progress?

Given the somewhat grumpy complaints of the previous section, it will not surprise when I suggest that the main problems in normative ethics remain much the same from period to period, however much we might repackage them. I have mentioned especially the agent-neutral/agent-relative divide between consequentialists and deontologists, but other examples readily come to mind in axiology (monism v. pluralism, subjective v. objective theories of the good, hedonists v. perfectionists, etc.). So the question is: if the problems are perennial, what are the prospects for progress with them? Are we getting somewhere or just recycling the past? Are we likely to get further with these issues in the future?

Perhaps it is just the cynicism of advancing age, but I have become more pessimistic about the likelihood that future generations of philosophers will do much better with these issues than we have. Forty years of concentrated effort by the best philosophical minds of our generation do not seem to have brought us appreciably closer to resolution. Instead, the same theories continue to have passionate advocates and equally passionate critics, with no prospect that any of these camps will be able to drive any other from the field. Perhaps a century hence some really smart whippersnappers will have this all figured out, but I doubt it. Furthermore, I have to confess that I don't want them to figure it all out, since that threatens to condemn my own labours of these past decades to obsolescence. A century hence, if I am lucky, it will be my work that will be coming back into view after having passed through its own dark age, and I want to think that it will still be worth reading.

### Bibliography

Broad, C.D., "Certain Features in Moore's Ethical Doctrines", in Paul Arthur Schilpp (ed.), *The Philosophy of G.E. Moore* (La Salle, Illinois: Open Court, 1942), pp. 43-67.

Greene, J.D., et al., "An fMRI Study of Emotional Engagement in Moral Judgment", *Science* 293 (2001): 2105-2108.

Hare, R.M., *The Language of Morals* (Oxford: Clarendon Press, 1952).

Hodgson, D.H., *Consequences of Utilitarianism* (Oxford: Clarendon Press, 1967).

Koenigs, Michael, et al., "Damage to the Prefrontal Cortex Increases Utilitarian Moral Judgements", *Nature*, online edition, doi:10.1038/nature05631 (21 March 2007).

Lyons, David, *Forms and Limits of Utilitarianism* (Oxford: Clarendon Press, 1965).

Sumner, L.W., *Abortion and Moral Theory* (Princeton: Princeton University Press, 1981).

— *The Moral Foundation of Rights* (Oxford: Clarendon Press, 1987).

— *Welfare, Happiness, and Ethics* (Oxford: Clarendon Press, 1996).

— *The Hateful and the Obscene* (Toronto: University of Toronto Press, 2004).

# 16

# Torbjörn Tännsjö

Kristian Claeson Professor of
Practical Philosophy
Stockholm University, Sweden
Affiliated Professor of Medical Ethics
Karolinska Institutet, Sweden

**Why were you initially drawn to normative ethics?**

When I was in my late teens I was already interested in philosophy. But the interest was theoretical. I wanted to know what it meant to know something, whether we can know something or not and, if so, on the basis of what kind of understanding. However, two simultaneous personal experiences drew me to *moral* philosophy. I was conscripted to military service, and my gut feeling was to refuse to serve. I did not want to kill other people. This seemed to me wrong, if not in principle, at least so in practice. There were no serious military threats facing Sweden, and if the situation were to change there would be no guarantee that I would serve among the good guys rather than among the bad ones. Moreover, the kind of values for which I was supposed to kill, such as democracy and national independence, were better served, I thought, through non-violent global political action. This was during the heyday of the civil rights movement in the American South. My arguments were met with no sympathy from the military authorities. They threatened me with jail, if I was not prepared to serve.

At the same time, when I was arguing with the military authorities, my father became ill. His illness soon turned out to be serious. He was suffering from cancer in his liver with many metastases elsewhere. The prediction was that he would be dead within a few months. This diagnosis was born out by realities. My father reacted with good sense and courage to the prophecy. He was sad to leave in such an untimely manner, he told my mother and me, but he had had more than fifty rich years, so he wasn't resentful. And he swiftly took care of all the practical matters relating to his death. However, something he had not expected occured. His sufferings turned out to be unbearable. He knew the doctor who treated him, and he was given morphine and all sorts of palliation. However, his pain did not respond. His last weeks were terrible. Sometimes he fell asleep and when he awoke he would still be in a delirious state caused by the morphine; he would ask me and my mother whether he was dead or alive. We had to tell him he had to struggle on for a little longer. He begged his doctor to assist him in his dying; asking for euthanasia. His doctor turned down his request with the words that euthanasia was not only illegal, but it was "at variance with the principles of medical ethics". My father's agony increased and culminated in a state of terminal agitation that ended only with his very last breath.

I was much concerned with what had happened. It not only affected me emotionally, I was intellectually in a state of deep confusion. How could it be that I had a legal obligation to kill people I did not know, and who had certainly not consent to being killed, while my father's doctor could not help my father to die when asked? My consternation brought me to moral philosophy and a life-long search for an answer to the question as to when and why we should, and when we should not kill. I began to study moral philosophy, or practical philosophy, as it was called at Stockholm University. However, the subject looked very different to what I had expected. And it was in no way obvious that there was a place for normative ethics in moral philosophy. Problems to do with the ethics of killing were not what at first confronted me.

**What example(s) from your work (or the work of others) illustrates the role that normative ethics ought to play in moral philosophy?**

In the 1960s Swedish philosophy was still very much under the influence of the late Axel Hägerström (1868–1939), Professor of Practical Philosophy at Uppsala University, and his "moral nihilism". Hägerström had rejected the idea that there are moral propositions that are absolutely true, independently of our conceptualization or thinking. When we pass moral judgements, according to Hägerström, we merely express our emotions. Now, since at the University, we ought critically and systematically to search for the truth, and since there is no truth to be found in morality, normative ethics should not be any part of the academic curriculum. Moral philosophy should be a study of, not *in*, morals, Hägerström had thus cautioned we moral philosophers.

All this meant that serious moral philosophy, when I took up my study, was metaethics (of a semantic orientation), period. So the important authors we read were thinkers like C.L. Stevenson, J.O. Urmson, and Richard Hare. And those who were interested in formalism investigated the logic of imperatives. Of course, G.E. Moore and Henry Sidgwick were read as well, but merely together with Plato and Aristotle, i.e. from a historical point of view. Their moral realism was considered as hopelessly dated. So at first I saw little possibility in pursuing the kind of moral philosophy in which I had become interested. My dissertation, consequently, was in metaethics ( *The Relevance of Metaethics to Ethics*, 1976).

However, my professor, Harald Ofstad, held a strong interest in normative ethics. He attacked utilitarianism and defended a kind of intuistic or particularist moral philosophy. He wanted practical philosophy to be both empirical and applied. He did not care about Hägerström's caution! And many philosophers in those days shamelessly started doing normative ethics, irrespective of what Hägerström and others had claimed about the impossibility of doing so in a scholarly manner. So reluctantly I followed suit.

It is true that many who discussed utilitarianism in those days were not themselves utilitarians. They held a *theoretical* interest in the doctrine and wanted to make it more comprehensible and precise, without committing themselves to it. Lars Bergström was a good example of this ( *The Alternative and Consequences of Utilitarianism*, 1966). My own interest in normative ethics started in a very detached manner. But internationally, many philosophers came in those days to *defend* strong normative positions, as Jack Smart and Richard Hare did with respect to utilitarianism, John Rawls with respect to his theory of justice, and Robert Nozick with respect to his theory of rights. All of a sudden normative ethics was "kosher" once again.

Does this mean that all moral philosophers had turned into moral realists? Obviously not. Some of them may have come to accept some kind of moral realism, but others stayed nihilists. Richard Hare, Jack Smart and Peter Singer, for example, defended utilitarianism in spite of the fact that they rejected the notion of an absolute moral truth to be found, rather than constructed, by us. And Rawls was vague on this point. However, these authors had in common the idea that there can be a distinction between bad and good arguments in moral philosophy. And they claimed that this distinction did not presuppose moral realism.

I must confess that I am still inclined here to side with Häger-ström. I find it difficult to make sense of notions such as consistency and logical consequence, or a reflective equilibrium, in an irrealist context (I write about this in "Understanding Through Explanation in Ethics", *Theoria*, 2006) and, even more importantly, I find it hard to see why we should *bother* with consistency and the related notions unless there is truth in the matter at hand. In a realistic context we want to avoid contradiction since we are interested in the truth and we know that at least one of two inconsistent propositions must be false, but there is no similar rationale for seeking consistency in an irrealist context.

These are difficult questions, of course, but I am prepared to defend the view that unless moral realism is true there is no place within moral philosophy for normative ethics. I suppose nowadays many philosophers are moral realists, but I suspect that some of them are as uncritical realists as people used to be uncritical moral nihilists in the 1960s. Hence, it is not satisfactory that the relation between moral metaphysics and normative ethics has not been made clearer.

I happen to believe that moral realism is true, so I feel comfortable when I pursue normative ethics. I would have hoped, however, that I could have given a better defence of moral realism than the one I have been able to come up with. I am utterly dissatisfied with the defence of moral realism I put it forward in *Moral Realism* (1990), but I am slightly more optimistic with some work I have in progress (a book ms, *Understanding Practical Reasons*).

So it is far from *obvious* that there is a place for normative ethics in moral philosophy. And, as if the problem with a threatening moral nihilism were not enough, there is another reason for thinking that normative ethics is problematic as well. The problem of free will should haunt moral philosophers. But my impression is that most of us do not bother with this problem.

I do not say that the distinction between, say, utilitarianism and Kantianism, makes no sense unless we have free will. Even if we do not have free will it might well be the case that utilitarianism is true and Kantianism false. However, the question of whether utilitarianism, or Kantianism, or any other normative theory, is true, has little *practical* importance, if we lack free will. And the reason that it has little practical importance has to do with Kant's observation that "ought" implies "can". If there are no alternatives to the actions we actually perform, if we cannot ever help doing what we do, then *all* our actions are deemed right by *all* competing normative theories, be they utilitarian, Kantian or of any other variety.

It seems to me that many normative ethicists are as complacent when it comes to the problem of free will as they are with respect to the problem of moral realism. In my *Hedonistic Utilitarianism* (1998) I devote a chapter to the problem of free will. I defend compatibilism. So once again I feel comfortable with doing normative ethics. However, the defence of compatibilism does not come easily.

As far as I can see, we cannot defend compatibilism unless we are prepared to countenance the possibility that we can sometimes perform actions such that, had we performed them, the past would have been different. This is no small concession to make, but unless we make it, there will be little practical point in doing normative ethics. And if we make it, we may have to acknowledge that we are responsible, not only for the future, but for the past as well! Or, perhaps there is some way of avoiding this astounding implication, but the problem has not yet received the attention it deserves.

It is no doubt that the philosophers are "back on the job", as Peter Singer once put it (*The New York Times Magazine*, 1974). He was referring to a revival of the interest in problems in normative and applied ethics. But *should* we be there? I think we should. But I am not quite sure as to how to *show* that this is so. If I am right, we should be there only if moral realism is true and we possess free will. These are big ifs indeed.

## What is the proper role of normative ethics in relation to the other disciplines?

Unless moral realism is true, there is no hope of progress in normative ethics. But moral realism is no guarantee of progress. Even if there are moral truths, existing independently of our conceptualisation and thought, it is an open question whether we can ever *find* them. Are there methods of investigation available in moral philosophy, similar to the ones we use in science, that render plausible the notion that we can gain moral knowledge?

I think there are. As a matter of fact, I think the methods are roughly the very same ones that we use in science. So even if Hume is right, that we cannot derive an "ought" from an "is", there is much for philosophers to learn from science from a *methodological* point of view.

Normative ethics has often in the past been conceived of in a foundationalist manner. The idea has been that there are some self-evident moral axioms that can provide a secure foundation for all our moral knowledge. This kind of approach has fallen into well-deserved disrepute nowadays. However, instead of it we often see a much too "intuistic" or "casuistic" approach to moral problems. Even very clever thinkers, whom I admire, sometimes make this mistake. These thinkers start out with examples, try to articulate principles that cater for our intuitions about the examples, apply the principles to other examples, find recalcitrant evidence, modify the principles, and move on to yet other examples ... while taking their intuitions for granted. The result is complication and confusion. I feel that this empiricist or intuitive or casuistic approach is just as hopeless in normative ethics as it is in science.

Just as we do in science, we ought in normative ethics to construct bold conjectures, which we test in many cases. But we should be prepared to *modify* our considered intuitions in the light of theoretical and other constraints.

How do we test the conjectures? We test them in the same manner that we test empirical hypotheses. We derive conclusions from them. If the conclusions we derive (to the effect that certain actions are right or wrong) are morally acceptable, then we may say that the moral conjecture in question explains the conclusions (it explains *morally* the rightness or wrongness of these actions). We then tend to speak of the content of our conjecture as a true representation of a moral *law*. This is parallel to when, in science, we tend to speak of our corroborated or inductively supported hypotheses as descriptive of *laws of nature*.

If a certain conjecture gives the *best* explanation of the rightness or wrongness of an action (assessed in terms of generality, adequacy, and simplicity, independent credibility, and so forth), then we may say that we have *evidence* for the conjecture. In that case we have made an inference to the best (moral) explanation. The existence of these explanatory and evidential relations renders our beliefs as being (more) justified.

This is no different from how we test empirical hypotheses in the sciences, then, with one important caveat. In morality, there is no observational basis of the testing. Instead we rely on our moral *intuitions*. However, just as we may, in the light of what observations we make, revise our empirical conjectures, we may revise our moral conjectures in the light of our intuitions, in order to arrive at more plausible moral theories. The process works the other way round as well. We sometimes find that we have to reconsider, even some of our more considered moral intuitions, just as we have to dismiss some seeming observational evidence in the sciences. In the interest of overall adequacy, simplicity and coherence, some intuitions or putative observations have to yield. But when they have to yield, we want to find an explanation for our mistake.

Or is this much too simplistic? Well, I have to admit that there is also another crucial difference between the sciences and morality. In the sciences we sometimes rely on thought-experiments, but this is the exception rather than the rule. In morality it is the other way round: thought-experiments are the rule and confrontation of our theories, with actual cases the exception.

This difference between the sciences and morality has to do with the fact that all plausible moral hypotheses are extremely demanding when it comes to collateral factual input. There is no way that we can put, say, Utilitarianism to test by "applying" it to a real case in order to assess whether the recommendations we deduce from it are intuitively acceptable. This is impossible since there is no way to deduce any definite recommendations of the kind we are after. There is no way of telling whether, from the point of view of Utilitarianism, a certain concrete action, performed by an agent in a concrete situation, was right or wrong.

All this explains why we need so desperately to have recourse to thought-experiments when we pursue normative ethics. In a thought-experiment we may safely abstract from difficult empirical questions. We assume that we know all the relevant consequences of an action.

To return to the problem that brought me to normative ethics, that is, the problem of killing. I have applied this approach to this problem. There are many ethical theories that have implications for the problem of killing, of course, but three of them stand out as being more influential and important than all the others: The Sanctity-of-Life-doctrine, the Right-to-Life-doctrine, and Utilitarianism. I am struck by the superiority of the utilitarian ethics of killing in comparison with the main competitors. I have carefully run my argument trough all sorts of killing, such as murder, abortion, euthanasia, capital punishment, and the killing in war, and I must say that I am satisfied with the utilitarian answers to the problems. And I am amazed at finding that I seem to be almost alone in thinking so. Even people like Peter Singer and Jonathan Glover, who sympathize with utilitarianism to some extent, hesitate to consequently apply it to the problem of killing.

According to Utilitarianism we ought to kill if, and only if, the consequences of killing are better than the consequences of not killing. It is difficult to know when, according to Utilitarianism, one ought to kill, and when one ought not to, but it is not far-fetched to believe that in many cases it is all right to kill. This may seem shocking, especially if we consider that there are many murders we should commit, but which we omit committing. But there is also another aspect to the utilitarian theory of killing. According to Utilitarianism, we should have those, and only those laws regulating killing that produce the best consequences. This prepares for a kind of double standard. In order to ascertain that we can sleep safely at night, we have to establish laws prohibiting, not only immoral killing, but all sorts of killing, at least if it can be properly described, in legal terms, as "murder" or "manslaughter."

I have to this date published only in Swedish, German, and Norwegian my investigations into the ethics of killing (the title of the book translated into English would be: *Thou Shalt Sometimes Kill*). The project is a bit slow, since I also am gathering empirical information about how people in different countries view different kinds of killing such as murder, abortion, euthanasia, capital punishment, and the killing in wars; my intention, however, is to carry this project further and to publish, in due time, an English version of the book (based on statistics from the US, China, Japan and Russia as well as from the countries I have investigated thus far).

## What do you consider the most important and yet neglected topics and/or contributions in normative ethics?

In my late teens, I was not only interested in the ethics of killing. My conscientious objection to the draft was inspired by an interest in global affairs. Hence in the late 1960s I joined a political organisation for the first time in my life. I became a Citizen of the World. My political hero was Garry Davies, who had burnt his passport, and who had claimed to be just that: a Citizen of the World. This is how he tells us about this:

... in May 1948 I renounced my US nationality at the Embassy in Paris, publicly declaring myself 'Citizen of the World'. With no national documents, I was considered stateless by France and ordered to leave by 12 September or be detained in jail. Meanwhile, however, the United Nations was preparing its 1948 session in Paris at the Palais de Chaillot which was, for the occasion, declared an 'international territory'. Unable to enter another country I 'entered' the new 'international territory' on the morning of 12 September and claimed 'global political asylum'. Many people got to hear of this through the international media and gave their support. A council of international intellectuals was formed, led by Albert Camus.[1]

Davies' organisation was still strong in the 1960s, but it did not last. My interest in global affairs, however, has remained and is now more intense than ever. As a matter of fact, my most recent book presents a strong defence of global democracy (*Global Democracy: The Case for a World Government* (Edinburgh: Edinburgh UP, 2008)). Global questions are not only interesting from the point of view of political philosophy. They are also of interest to normative ethics proper. As a matter of fact, global affairs seem to raise *special* problems in normative ethics.

A common observation, when global problems are discussed, is that they are so huge, so formidable, that it makes no sense individually to bother with them. I think there is some truth in that kind of comment, but I still feel that there is no more urgent a task than for an individual in today's world than to bother with wars, global injustices and global environmental problems; so there seems to be a theoretical problem facing us here.

In order to get at a clear diagnosis of the problem, we need to distinguish between two kinds of cases. In one case, we face *threshold* effects. Then it is literally true that, irrespective of whether I make a contribution to a joint cause or not, I will make no difference. In the other kind of case, I do make a difference, but it is so tiny that it threatens to go unnoticed.

Thinkers such as Jonathan Glover and Derek Parfit have drawn our attention to these kinds of problems. I have been fascinated with both kinds, and I believe the best way of treating them is along the following lines.

When it comes to threshold effects, we need to ponder the question of collective responsibility. The idea is not that an individual is responsible for what a collectivity has done, nor is it that a collectivity is responsible for what certain individuals have done, but rather that the collectivity as such bears responsibility for what the collectivity has done (or, more typically, has omitted to do). In my *Hedonistic Utilitarianism* I defend a version of utilitarianism

---

[1] Accessed on November 6, 2005: http:// www.newint.org / issue289 / xword.htm

where I acknowledge that collectivities can act, can act wrongly, and that they bear responsibility for their actions. A nice theoretical gain with this position is that, once we admit that collectivities can act, we obtain the result that, if *all* agents, including the collective ones, act in accordance with the demands of utilitarianism, then the result is optimal. I have also recently speculated about the political implications of this view in my "The Myth of Innocence. On Collective Responsibility and Collective Punishment".

When the effects seem negligible, on being assessed from the point of view of each individual who is affected by what I do, it might still be the case that the *sum total* of the disvalue I produce is considerable. This is so if very *many* people are affected by what I do.

It seems to me that unless we acknowledge that very slight effects, even sub-noticeable effects, are of moral importance, we cannot handle some of the most pressing global problems to do, say, with global warming.

I have developed this line of thought, which is an elaboration of ideas already put forward by Bentham and Edgeworth, in their search for a hedonistic atom (the least noticeable difference with respect to well-being). One can perhaps say that I go even further then they do, when I speculate about a possible *fission* of this atom into sub-noticeable (or merely indirectly noticeable) differences with respect to well-being. This is not merely a way of making sense of hedonistic utilitarianism and to try to solve the problem with interpersonal comparisons of well-being, but also a theme of direct relevance to the solution of theoretical moral problems engendered by existing global problems.

I do not claim that the problems to do with threshold effects and sub-noticeable differences of well-being are entirely neglected in the present philosophical discussion, but I do feel that they are so urgent, that one should expect more interest to be vested in them.

There is a third problem, which has fascinated me, which is also directly connected to global political questions: population ethics. Some thinkers hold that it doesn't matter if there are sentient creatures around some thousand years hence. Others have held that there is no task more urgent than to see to it that there are. And some have gone to great lengths and argued, like the present author, that the more sentient beings there are, the better (if their level of well-being is the same).

A standard objection to arguments like the one I have put for-

ward is that they lead to what Derek Parfit has nicknamed the
"repugnant" conclusion, that we should prefer an enormous population living lives barely worth living to a more restricted population (of ten billions, say), leading extremely happy lives. This
observation is correct. It seems impossible to avoid this conclusion,
once you adopt, in my view, a decent view of global problems and
world politics. But then, I do not find the conclusion at all repugnant.

I think when people find it repugnant they make a moral mistake, which is structurally similar to the mistake we make when
we claim that it doesn't matter whether we contribute or not to
pollution. We fail to grasp the importance of large sums made up
by a greate many very tiny numbers. We fail to acknowledge the
moral importance of the sum-total of many tiny contributions to
a whole of well-being. If this is a mistake with respect to the effects of our individual bad behaviour towards global warming, it is
probably also a mistake when we fail to acknowledge the value of
an immense population, where each individual leads a life worth
living.

This is not the place to argue one way or another with respect to
the repugnant conclusion. I have done so elsewhere (in my paper:
"Why We Ought to Accept the Repugnant Conclusion".) However, there can be no denying that problems in population ethics
are of the utmost relevance to global politics. And yet they seem
to trouble only a very restricted group of moral philosophers. The
rest fail to take them seriously. Why is this so?

To me this is a mystery. Over and over again I have tried in
vain to educate, not only my fellow moral philosophers about the
political relevance of questions in population ethics, but ordinary
lay people as well. Both tend do diagnose these problems as examples of sophistry. They are wrong but I don't know how to
communicate this to them!

## What are the prospects for progress?

I believe there is a truth in (normative) ethics to be found, rather
than constructed, and I believe there exist methods of investigation rendering it possible systematically and critically to pursue
the truth. So there should be room for some optimism with respect to the future of normative ethics. However, the one problem
that has always plagued moral philosophy is religion. People have
not been allowed to think freely about moral problems. This has

posed an enormous obstacle to progress in moral thinking. And it is hard to conjecture what the future holds.

I suppose that, had I lived during the enlightenment, I would have been very optimistic. When there is progress in the sciences people will become secularised. And when dogmatic religion goes away, space will open up for a free pursuit of moral truth.

However, I happen to live in a post-modern world, where it is difficult to feel any confidence about the future. I cannot help to think, that many people have some kind of religious need, perhaps they want to stand on ceremony and yield to authorities rather than think critically. If this is the case, moral philosophy in general, and normative ethics in particular, will remain a field for a few dedicated experts. This would be bad because the isolation of the subject from people at large and from their concerns threatens to leave important aspects of it in the dark.

# 17

# Larry S. Temkin

## Professor of Philosophy

Rutgers, The State University of New Jersey, USA

---

**Why were you initially drawn to normative ethics?**

I was initially drawn to normative ethics because I believed that ethics was fundamentally important; indeed, the *most* important discipline a philosopher might engage in. I still believe that. To be sure, progress in ethics is slow, and hardly inevitable. But it can be made. Moreover, importantly, there is great value in attempting to make such progress, even if our efforts eventually fail. Indeed, here, as elsewhere, the distinctively human and ennobling task of asking and pursuing the right questions may itself be as valuable as any answers we may find.

Of course, the question of how I came to be the sort of person who believed that ethics was fundamentally important is another matter. Mostly, I trace it to the crucial influence of my parents and grandmother, to my natural temperament, to important events that were going on while I was growing up, and to luck. Let me comment on each.

### Family influences

My father, Blair Temkin, entered college at age sixteen, and the army two years later. An infantryman in World War II, he never discussed the war; but as an adult I learned, via an old letter, that he was the only member of his platoon to survive the assault on Okinawa. After the war my father returned to school, earning a BS and MS in chemical engineering. My father had been an Eagle Scout, and earned Scouting's highest honor, the Order of the Arrow. More than anyone I ever knew he embodied the Scout Code. He really *was* "trustworthy, loyal, helpful, friendly, courteous, kind, obedient, cheerful, thrifty, brave, clean and reverent." Putting his family above all else, he worked long, grueling, hours as a foundryman, often seven days a week. He eventually became a successful businessman, but he still worked six days a week. A conservative, patriotic, humanitarian, my father was a font of stories conveying moral and life lessons. He served as a scout master for many years, and supported many charities. Especially significant for me were the letters and pictures he received conveying the impact his contributions had on needy overseas children. Thoroughly reasonable and evenhanded, my father taught us, by example, the importance of integrity, personal responsibility, leaving the world a better place, and treating all people fairly and respectfully.

My mother, Leah Temkin, married before finishing college. But with four young children she returned to school, finishing her BA, and then later her MA and Doctorate of Education. She served as president or vice-president in various local, regional, and district volunteer organizations, and headed up the Welfare Department's first program to teach reading to welfare recipients. From age 37 to 55, my mother worked full time teaching adult reading, mostly to the poor, immigrants, and people of color. A compassionate, concerned, optimistic liberal, she, too, stressed the importance of treating everyone respectfully, and generously supported many charities. By her example, we saw the possibility of attaining self-realization, without ignoring the needy.

My grandmother, Faye Sigman, was a strong woman who publicly dressed down Senator Joseph McCarthy at the height of his anticommunist witch hunt. Founder of the first synagogue in her community, she impressed on me the importance of my Jewish heritage, and the lessons to be learned from it. The persecution of Jews from Roman times, through the Inquisition, through the Holocaust made vivid the dangers of fear, prejudice, ignorance, hatred, and intolerance.

Together, these three people nurtured a firm sense of right and wrong, and a keen sense of injustice and unfairness. They led me to respect each person, and taught me that decent people could often *reasonably* disagree even about fundamental issues. Thanks to them, I saw the importance of self-fulfillment, and giving special weight to family and friends, but I also learned the importance of giving weight to the interests of others. I was acutely aware that no one *deserved* a better start in life than others, and that our obligations extended well beyond those closest to us. I learned, as Rabbi Hillel put it in the *Sayings of the Fathers*, "If I am not for me, then who will be? If I am only for me, then what am I?"

## Natural temperament

As long as I can remember I have cared about ethical matters. When children taunted other, weaker, children, it deeply bothered me. When, at 10, I saw a black woman begging on a blanket surrounded by young children, I burned with guilt, shame, and impotence. Comparing their situation with mine, it seemed clearly unfair and deeply wrong. I experienced such feelings frequently as a youth, and they have persisted. Whenever I contemplate those who have unfairly suffered, whether they are poor, handicapped, women, minorities, or homosexuals; or victims of famine, disease, accident, catastrophe, or war; the old feelings of guilt, shame and impotence return. Or, rather, more sophisticated "adult" versions of those emotions arise. No doubt such "reactive attitudes" helped spur me to do normative ethics, as I sought to better understand why we *should* care about situations, and what we should *do* about them.

## Important Events

Many important events helped forge my concern about ethics. At seven, I helped stock supplies in our basement during the 1961 Cuban Missile Crisis. At nine, I watched the burial of President Kennedy following his assassination. At ten, our school ran civil defense drills in case of nuclear attack. In 1967, when I was 13, the Civil Rights Movement erupted into rioting across the United States. In 1968, Martin Luther King Jr. and Bobby Kennedy were assassinated. That same year, there were violent demonstrations against the Vietnam War, my favorite radio station ended every newscast with the words "and the war continues," and radical groups like the SDS, the Weathermen, and the Black Panthers engaged in armed struggle against the government. Paul Erlach published his apocalyptic vision about the dangers of overpopulation, *The Population Bomb*.[1] At the same time, feminists were becoming increasingly vocal, as were environmentalists. In 1970, when I was 16, four Kent State students were killed by National Guardsmen. Four years later, millions died in Bangladesh's great famine.

These events, and others, seared my consciousness. Great battles were raging about issues of fundamental moral importance, and people were deeply divided. Clearly, people needed direction.

---

[1] Ballantine Books, 1968.

Moral direction. In such circumstances, normative ethics had great appeal. Might it provide us with precisely the direction so desperately needed?

However unrealistic, it was an appealing thought. And it still is.

## Luck

In many ways, it is a matter of luck that I even became a philosopher. But it is especially lucky that I became a philosopher when I did. As an undergraduate, I was keenly interested in ethics. But I was warned that no "serious" philosopher did ethics; hence, I should reserve ethical concerns for my personal life, and focus on "respectable" topics like philosophy of language or logic. And so I did.

Fortunately, I arrived at Princeton for graduate school just one year after John Rawls published his great work *A Theory of Justice*. Rawls's book "legitimized" the pursuit of value theory, and enabled me, and many of my peers, to focus on ethics. It was an exciting time for value theory, and I was fortunate to take courses with Tom Nagel, Tim Scanlon, Gil Harman, and Derek Parfit. They showed that normative ethics could be serious, rigorous, interesting, and important. I have been lucky to do normative ethics, and political philosophy, ever since.

## What examples from your work illustrate the role that normative ethics ought to play in moral philosophy?

Normative ethics may be practical or theoretical. It may address particular normative issues, like the permissibility of capital punishment or our obligations to the needy. It may focus on different objects of moral evaluation, for example, outcomes, agents, or actions It may explore the nature and significance of particular notions, like equality or rights. It may also assess overall moral theories, like deontology, consequentialism, or contractualism.

Many have firm views about the "big" ethical issues. But few reflect on their views to determine if they withstand scrutiny. Normative ethics should explore the nature and justification of our moral beliefs. We need to examine our views' foundations, so as to understand their implications, and whether they are compatible. We also need to consider the strongest arguments for and against our views, to determine if they are rationally defensible or in need of revision. Ultimately, we seek a moral theory that enables us to justify as many of our deepest moral intuitions or beliefs as possible, in a clear, coherent, plausible, systematic, and non-ad hoc way, while illuminating which of our pre-theoretical views should be revised or rejected.[2]

My own work has focused on three main areas: egalitarianism, the good, and obligations to the needy. Let me comment on each.

Many believe that equality is valuable; but examination reveals deep disagreement about *why* equality matters. For *instrumental* egalitarians, equality is valuable only insofar as it promotes other ideals, such as freedom or utility. For *non-instrumental* egalitarians, equality is sometimes valuable in itself, beyond the extent to which it promotes other ideals. For some (*deontic*) egalitarians, people have a duty to promote equality in certain circumstances. For other (*telic*) egalitarians, equality is a feature of the goodness of outcomes. For some, inequality is only bad when it is bad *for* people. For others, inequality – like injustice or acting wrongly – can be bad, even if there is no one for whom it is bad.

Many positions are often conflated with egalitarianism. For example, I have argued that many who think of themselves as egalitarians are best regarded as *prioritarians*. For prioritarians, the worse off people are, the more priority should be given to improving their positions. So, like egalitarians, prioritarians will favor increasing a worse off person by a certain amount rather than a better off person by the same amount, and will typically favor transfers from the better to worse off. Still, prioritarians don't care about equality *per se*. Thus, prioritarians would see no objection to raising the best off a lot, if this were necessary for raising the worst off a little. Indeed, prioritarians would approve raising the best off a lot, even if the worst gained nothing, or were even lowered some, though this would greatly increase the outcome's inequality.

Many assume that the notion of equality is simple, holistic, and essentially distributive. They assume it is *simple*, because they think that we *all* know what equality is, that is just everybody having the same amount of x for whatever x we are interested in; *holistic*, because people often focus on inequalities between different *groups*, for example, blacks and whites, or men and women;

---

[2] This approach to ethics was championed by Henry Sidgwick, in *The Method of Ethics*, $7^{th}$ edition, 1907, and it has many adherents, including John Rawls, Derek Parfit, Frances Kamm, Shelly Kagan, Jeff McMahan, and Thomas Hurka.

and essentially *distributive*, because the ideal of equality seems to express a concern about how goods are distributed amongst different groups. After exploring egalitarianism's roots, I argued that this conception of equality is deeply misleading. Specifically, I argued that a fundamental notion of equality is intimately concerned with fairness, and that this notion is enormously *complex*. In particular, I argued that there are at least twelve different aspects underlying our egalitarian judgments. Additionally, I argued that this notion is *individualistic*; so that if, in fact, inequality mattered between rich whites and poor blacks, it would also matter between *any* rich and poor individuals, including, for example, rich and poor blacks, or rich blacks and poor whites. Finally, I argued that this notion of equality is essentially *comparative*; after all, equality is a relation, and what is *distinctive* about egalitarianism is its concern for how individuals fare *relative to each other*.

Recognizing that the concern for equality is ultimately connected with a concern about *comparative fairness* helps explain the sense in which one might care about equality beyond the extent to which it is good or bad *for* people. This is because a concern about fairness or justice is not simply reducible to a concern about how people *fare*. Thus, it is plausible to believe that an outcome in which sinners fare well is bad in *one* important respect, because "absolutely" unfair or unjust, even if it isn't bad *for* the sinners, or anyone else, that the sinners fare well. Similarly it is plausible to believe that an outcome where some are better off than others no less deserving than they, is bad in *one* important respect, because *comparatively* unfair, even if it isn't bad *for* anyone that the better off people are faring better. So, for example, if women receive what they deserve for their work, but men receive much *more* than they deserve for the same work, that may be bad, because unfair, even if it isn't any worse *for* the women (or the men) that the men have received more. This position helps answer the most powerful anti-egalitarian objection, *the leveling down objection*, which claims that equality doesn't matter, since there can never be *anything* good about promoting equality *merely* by leveling down the better off to the level of the worse off. The egalitarian can plausibly reply that leveling down *can* be better in *one* important respect, namely, when it is better regarding *comparative fairness*, even if she grants that it *may* not be better *all things considered*. For the egalitarian, fairness matters, and in particular comparative fairness matters, though it is not the *only* thing that matters.

Countless social, political, and moral arguments are couched in egalitarian terms. But, until one fully understands the notion of inequality, assessing such arguments is hard. Exploring the notion of inequality reveals insights about its nature, roots, complexity, and implications, enabling one to better evaluate egalitarian and anti-egalitarian claims. This is an important role normative ethics can play.[3]

Consider, next, the nature of the good. John Rawls once wrote that "All ethical doctrines worth our attention take consequences into account in judging rightness. One which did not would simply be irrational, crazy."[4] Surely, Rawls is right. Consequences often matter. Moreover, often, we either need or want to determine the best outcome, "all things considered." Thus, echoing Rawls, all plausible theories of practical reasoning must give an adequate account of the assessment and ranking of outcomes. Unfortunately, an adequate account has not yet been given; nor is one on the horizon. In exploring how we evaluate outcomes, I have discovered that some of our deepest views on this topic are incompatible. Let me illustrate some of the considerations relevant to this issue.

Most people accept a position I call the *First Standard View*, (*FSV*), which holds, roughly, that an outcome in which some people suffer a burden would be better than an outcome where many more people suffer a slightly smaller burden. So, for example, other things equal, an outcome where some number of people suffered from a given illness would be better than one where far more people suffered from an illness that was almost as bad. Most people also accept a position I call the *Second Standard View*, (*SSV*), which holds, roughly, that an outcome where a number of people suffer a really major burden would be worse than one where any number of people suffer a minor burden. For example, most be-

---

[3] For a more extended overview of my egalitarian views see my contribution to the companion volume *Political Questions: 5 Questions on Political Theory*, edited by Morten Ebbe Juul Nielsen, pp. 147-167, Automatic Press/VIP, 2006. For a deeper exploration of my egalitarian views see *Inequality*, Oxford University Press, 1993; "Egalitarianism Defended," *Ethics* 113, no. 4, pp. 764-782, 2003; "Equality, Priority, or What?" *Economics and Philosophy* 19, no. 1, pp. 61-88, 2003; "Egalitarianism: A Complex, Individualistic, and Comparative Notion," in *Philosophical Issues*, volume 11, eds. Sosa, Ernie and Villanueva, Enriquea, pp. 327-352, Blackwell Publishers, 2001; and "Equality, Priority, and the Levelling Down Objection," in *The Ideal of Equality*, eds. Clayton, Matthew and Williams, Andrew, pp. 126-161, Macmillan and St. Martin's Press, 2000.

[4] *A Theory of Justice*, Harvard University Press, p. 30, 1971.

lieve that, other things equal, 10 people suffering from quadriplegia would be worse than any number of people suffering from a brief, mild, headache. Finally, most people also accept an *Axiom of Transitivity*, which holds that if, all things considered, A is better than B, and B better than C, then A is better than C.

Unfortunately, these three, deeply held, beliefs are incompatible if, as seems possible, there could be a spectrum of burdens ranging from the very severe to the very mild, such that FSV applies when comparing outcomes involving burdens near each other on the spectrum, while SSV applies when comparing outcomes involving burdens at opposite ends of the spectrum. Thus, FSV tells us that outcome A, where 10 people are quadriplegic, would be better than outcome B, where 30 people suffer a burden almost as bad; that B would be better than outcome C, where 90 people suffer a burden almost as bad as the burden suffered in B, and so on. Continuing in this way, transitivity will entail that A, an outcome where 10 people are quadriplegic, is better than some outcome Z, where many people suffer a brief mild headache. But SSV denies this. Thus, one must reject FSV, SSV, or the Axiom of Transitivity; but none of these will be easy to give up.

FSV reflects the fact that *sometimes* we adopt an *additive aggregationist approach* in evaluating and comparing different outcomes. That is, we judge the relative goodness of two outcomes by comparing them in terms of both the quality *and* number of benefits or burdens and adding them up. SSV reflects the fact that *sometimes* we adopt an *anti-additive aggregationist approach* in evaluating and comparing different outcomes. That is, for *some* comparisons, we don't simply *add up* the benefits and burdens in the different outcomes, but instead pay attention to how the benefits or burdens are distributed in the different outcomes and, in particular, to the relative impact on people's lives that the benefits and burdens have. But I've shown that *if* we apply one set of criteria for making certain comparisons, and another set for making others, then the Axiom of Transitivity will either fail, or fail to apply, across the different comparisons. In particular, if FSV were relevant for comparing A with B, and B with C, but SSV were relevant for comparing A with C, then it won't be surprising if A is better than B, and B better than C, in terms of the relevant criteria for making *those* comparisons, but A is *not* better than C, in terms of the relevant criteria for making *that* comparison.

Similar considerations apply when thinking about different possible distributions of benefits and burdens within a life. Most believe that an analogue of FSV is relevant for comparing certain possible lives, but that an analogue of SSV is relevant for comparing others. For example, in some cases we accept an additive aggregationist approach, agreeing that it would be better to live a life with a larger burden lasting a certain duration, than a life with a burden almost as bad lasting much longer. But in other cases we reject an additive aggregationist approach, holding, for example, that no matter *how* long we might live, it would be better to have one extra mosquito bite per month for any number of months, than two years of excruciating torture during our life. Two years of torture within a life is tragic. But lots of mosquito bites spread out through time never amount to more than a nuisance; they simply don't *add up* in the way they would need to outweigh the tragic impact of two years of torture.[5] But once again, these views, which almost everyone accepts, entail that the Axiom of Transitivity either fails, or fails to apply, across comparisons of different possible lives.

FSV and SSV are restricted in scope. Each seems relevant for making certain comparisons, but not others. In *A Theory of Justice* Rawls makes it plain that his two principles of justice are also restricted in scope. Thus, Rawls acknowledges that there are surely circumstances in which they fail (p. 63)," and specifically contends that his principles only apply where civilization is "sufficiently advanced." More generally, it is arguable that many principles are restricted in scope. For example, some principles seem plausible and relevant for comparing alternatives involving the same people, but deeply implausible and irrelevant for comparing alternatives involving different people. But if this is right, then once again different criteria will be relevant for comparing different outcomes, and the Axiom of Transitivity may either fail, or fail to apply.

Some people doubt that "better than" could ever be intransitive. However, most readily grant that since context is relevant to our *obligations*, it could be that we ought to do A rather than B, when those are our only alternatives, and do B rather than C,

---

[5] This example is a variation of one first sent to me by Stuart Rachels. His important work on this topic includes "Repugnance or Intransitivity," in *The Repugnant Conclusion: Essays on Population Ethics*, edited by Ryberg, Jesper and Tannsjo, Torbjorn, pp. 163-186, Kluwer Academic Publishers, 2004; "A Set of Solutions to Parfit's Problems," *Nous* 35, number 2, pp. 214-238, 2001; and "Counterexamples to the Transitivity of *Better Than*," *Australasian Journal of Philosophy* 76, no. 1, pp. 71-83, 1998.

when those are our only alternatives, and yet we ought to do C rather than A, when those are our only alternatives. But if the right is relevant to (even if not determinate of) the good, then it might *also* be the case that the outcome in which we do A would be better than the outcome in which we do B, when those are the only alternatives, the outcome in which we do B would be better than the outcome in which we do C, when those are the only alternatives, and yet the outcome in which we do C would be better than the outcome in which we do A, when those are the only alternatives. In this case transitivity would fail to apply across the different sets of outcomes, and hence we would be unable to use our pairwise judgments about the different outcomes to help determine which outcome would be best all things considered.

In sum, I believe that many of our deepest beliefs regarding how to assess the goodness of outcomes are fundamentally incompatible. Unfortunately, changing our beliefs has deeply implausible implications, and may require significant revision in our understanding of the good and the nature of practical reasoning.[6]

Let me next discuss obligations to the needy. I believe that many different kinds of considerations are morally significant, and hence that we must be moral pluralists in our moral deliberations. For example, normative ethics has to consider the goodness of outcomes. It also has to consider issues of virtue and vice, because character also matters. Likewise, it has to consider the issue of duties, to ourselves and others. Considering each of these areas, it seems clear that we are open to serious moral criticism if we ignore the needy.

For example, notwithstanding the serious worries expressed above,

---

[6] I am exploring these issues in a book length manuscript tentatively titled "Rethinking the Good, Moral Ideals, and the Nature of Practical Reasoning." For some of my publications on these topics see "A 'New' Principle of Aggregation," *Philosophical Issues*, 15, *Normativity*, edited by Sosa, Ernest and Villanueva, Enrique, pp. 218–234, 2005; "Worries about Continuity, Transitivity, Expected Utility Theory, and Practical Reasoning" in *Exploring Practical Philosophy*, eds. Egonsson, Dan, Josefsson Jonas, Petersson, Björn, and Rønnow-Rasmussen, Toni, pp. 95-108, Ashgate Publishing Limited, 2001; "An Abortion Argument and the Threat of Intransitivity," in *Well-being and Morality: Essays in Honour of James Griffin*, eds. Crisp, Roger and Hooker, Brad, pp. 336–356, Oxford University Press, 2000; "Rethinking the Good, Moral Ideals and the Nature of Practical Reasoning," in *Reading Parfit*, edited by Jonathan Dancy, pp. 290-344, Basil Blackwell, 1997; and "A Continuum Argument for Intransitivity," *Philosophy and Public Affairs* 25, no. 3, pp. 175–210, 1996.

surely on *any* plausible theory of the goodness of outcomes, an outcome where many innocents die of easily avoidable causes will be worse than one where affluent people spend less on cars, clothes, eating out, or recreational items. Likewise, among the most important virtues are beneficence, sympathy, compassion, and generosity. But then, if one takes being virtuous seriously, surely, at *some* point one must give priority to easily preventable hunger, illness, and suffering, over further acquisitions that one doesn't need, may hardly use, and wouldn't miss if one didn't have them. Finally, most people recognize that we have both positive and negative duties, and that some positive duties are more pressing and important than some negative duties. For example, the positive duty of saving a drowning child is far more important than the negative duty of not stealing a piece of candy. Now I believe there is a strong positive duty to relieve severe pain and suffering, whenever we could easily do so without significantly impacting our lives negatively. But, by that criteria, most affluent people act wrongly if they ignore the needy. For, surely, most affluent people *could* easily help the needy without suffering a significant negative impact.

Some people claim that the needy have no *right* to our assistance, that *acting unjustly* involves rights violations, and that acting *wrongly* involves acting unjustly.[7] Each claim is disputable, but even if one grants that one doesn't violate any rights, or *act* unjustly if one ignores the needy, there can still be powerful reasons *of justice* to aid the needy. For example, according to *absolute justice*, good people deserve to fare well, and it is unjust when they fare poorly; and according to *comparative justice* it is unjust for some to fare worse than others no more deserving. Thus, on these views there will be reasons *of justice* to aid needy people who don't deserve their absolute or comparative plight. In any event, rights violations are not the only factors that make an act wrong. My earlier suggestion stands. Even if we grant the controversial assumptions that the needy have no *right* to aid, and that we don't *act* unjustly if we ignore them, we are still open to serious moral criticism for ignoring the needy, when we could easily aid them at no significant cost to ourselves.[8]

---

[7] See, for example, Jan Narveson's "Welfare and Wealth, Poverty and Justice in Today's World," and "Is World Poverty A moral Problem for the Wealthy," in *The Journal of Ethics* 8, pp. 305–348 and 397–408, 2004.

[8] My views regarding obligations to the needy are further developed in "Thinking about the Needy, Justice, and International Organizations," and

## How do studies within scientific disciplines contribute to the development of normative ethics?

I believe that there are at least five ways in which scientific disciplines may influence the direction and impact of normative ethics.

First, scientific studies can be directly relevant for determining how best to pursue a moral goal. For example, many disciplines, including agronomy, animal husbandry, biology, nutritional science, medicine, pharmacology, economics, and political science might all be relevant in deciding how to effectively address world hunger. But, importantly, none of these disciplines will have much to say about whether alleviating hunger *should* be a moral goal, or how much weight we should give it relative to other goals. I shall return to this point below.

Second, scientific studies may illuminate the scope of a moral principle. For example, suppose we decide that killing innocent beings with certain psychological capacities is wrong. Advances in psychology, neurophysiology, and zoology might then reveal that killing certain animals is wrong, because they possess the capacities in question. Certainly, current attitudes towards dolphins and monkeys reflect scientific developments regarding the psychological capacities of these animals.

Third, scientific advances may cause us to abandon some moral beliefs. For example, advances in physics, psychology, genetics, neurophysiology, or pharmacology may convince us that some individuals are not responsible for particular behaviors. Perhaps we might discover a gene, or chemical imbalance, that fully accounts for certain anti-social behaviors. In that case, even if interventions were warranted, we might decide that people should not be blamed or held responsible for such actions. As before, this normative claim would be dictated by ethics, not science; but science would have revealed its applicability.

Fourth, scientific advances may prod us to reexamine the foundations of some moral beliefs. For example, it has long been recognized that humans should be treated with respect. Prior to evolutionary theory, many thought that this was because humans were created in God's image; afterwards, there was reason to seek another foundation for the view in question.

Fifth, scientific advances can illuminate our nature, and with it our moral predicament. For example, by illustrating how easily

"Thinking about the Needy: A Reprise," *Journal of Ethics* 8, pp. 349–395 and 409–458, 2004.

normal people can be induced to (seemingly) inflict great pain on others, Milgram's famous psychological experiments raise deep questions about people's characters, the foundations of evil, and moral luck.[9] Thus, such experiments buttress the suspicion that our characters may not be all that different from those of many Nazi collaborators, and that it may be largely a matter of good or bad luck who finds themselves in what positions, who does or does not participate in great evils, and who is (regarded as) virtuous or vicious.

Having said this, let me emphasize that I believe scientific disciplines provide little guidance in determining *fundamental* moral principles. For example, when scientists claim to offer a social, genetic, or evolutionary explanation for why humans are altruistic, or have certain moral sentiments, these accounts of morality are irrelevant to whether there is good *reason* to be altruistic, or have certain moral sentiments. That is, the normative question of whether we *should* be a certain way, or *should* care about certain things, is not settled by such scientific explanations. Similarly, I am not persuaded that MRI experiments, or any other experiments, have much bearing on what we *should* believe regarding trolley problems, or any other moral issue.[10] If we discovered that the brain's so-called "rational" part lit up in those Nazis who decided to put people in the ovens, but that the so-called "emotional" part lit up in those who refused to do so, that shouldn't alter our judgment that the former acted wrongly. Like all fundamental moral truths, the question of whether one should or shouldn't put people in the ovens is a normative one, and its answer doesn't depend on, and won't be revealed by, empirical scientific studies.

---

[9] See Stanley Milgram, *Obedience to Authority: an Experimental View*, HarperCollins, 1974.

[10] So-called trolley problems involve scenarios where one could save a larger number of people only by killing a smaller number of people, as, for example, by diverting a trolley from a track with five people onto another track with only one person. They raise a host of deep moral questions, and have been widely discussed in the philosophical literature. Joshua Greene and others have argued that we can learn important lessons about morality by studying the MRIs of people confronting trolley problems. See, for example, Greene, J.D., "From neural 'is' to moral 'ought': what are the moral implications of neuroscientific moral psychology?" *Nature Reviews Neuroscience*, vol. 4, pp. 847-850, 2003; and Greene, J.D., Sommerville, R.B., Nystrom, L.E., Darley, J.M., and Cohen, J.D., "An fMRI investigation of emotional engagement in moral judgment," *Science*, vol. 293, pp. 2105–2108, Sept. 14, 2001.

Moral principles are true in all possible worlds, and will often say something like this: In circumstance $C$, one ought to do $X$; or in circumstance $C$, outcome $A$ would be better than $B$. Science can tell us whether or not circumstance $C$ obtains, or whether our world is, in fact, like outcome $A$ or $B$, and this is crucially important if we are to correctly apply the moral principles. But science itself cannot discover the correct moral principles. This is the task of normative ethics.

## What do you consider the most neglected topics and/or contributions in normative ethics?

Within the public at large, virtually all of the writings of professional moral philosophers are neglected. The situation is only slightly better within philosophy itself. Indeed, even amongst ethicists, most people only work within their own small area of the ethical domain, and largely ignore the rest. Certainly, this is true of me. So, rather than trying to indicate the most neglected topics or contributions – which would involve a very large tie between lots of works! – let me make a small suggestion regarding some of the material most worth considering.

I am a firm believer that we stand on the shoulders of giants that we may see further. In normative ethics, most of the giants are dead. Although there is a lot of important work on contemporary normative issues, my own view is that anyone interested in pursuing normative ethics should have a firm grounding in normative theory. To do this, I would advise that before reading the latest books and journals devoted to ethics, one explore the writings of the philosophical giants who have profoundly shaped our understanding of ethics.

Among philosophers, I think it is most worth reading, and rereading, Plato, Aristotle, Butler, Hume, Kant, Mill, Nietzsche, and Sidgwick. I would also commend the great political theorists, Hobbes, Locke, Rousseau, Marx, and, perhaps, Machiavelli. Of course, I'm not suggesting that one has to read all of the writings of each of these figures. Nor am I trying to provide an exhaustive list of the major figures; no doubt others might also be listed, like Hegel, Bentham, and Adam Smith. I simply want to emphasize the importance of the major figures for normative ethics. Of course, one shouldn't stop with the historical giants. One should also read Moore and Ross, as well as more contemporary figures like Nagel, Nozick, Parfit, Rawls, Singer, and Williams, among others.

No doubt some scientists might find this amusing, and indicative of philosophy's backward state. After all, in many scientific domains it would be a waste of time, except for historical reasons, to consider the results of 1000, or 100, or even, in some cases, 20 years ago. But moral philosophy is not like science. Though we have made significant progress in normative thought, many of the insights of our philosophical predecessors are fundamentally important. These insights should not be overlooked.

## What are the most important problems in normative ethics and what are the prospects for progress?

Let me begin by noting some important problems in three different domains: applied, theoretical, and political. I'll then briefly discuss the prospects for progress.

### Applied

In thinking about the most important problems in applied ethics, it is hard not to be impressed by problems that affect large numbers of beings in major ways. Thus, amongst the most pressing ethical issues globally are poverty, hunger, health, education, and the treatment of women. Other issues at the forefront of ethical concern are global warming and other environmental problems, along with the related issues of economic development, sustainable growth, and population policies. Likewise, racism, ethnic cleansing, sectarian violence, and terrorism need to be addressed, especially when combined with the dangers of biological, chemical, and nuclear weapons proliferation. Genetic and medical enhancements have potential consequences too far reaching to be ignored. In addition, the treatment of animals is of enormous significance as it affects billions of sentient beings.

This is not to deny the urgency of other problems, including personal ethical dilemmas. But what is at stake with the above issues is so great, and affects so many in both present and future generations, that surely they are amongst the most important problems in applied ethics.

### Theoretical

As I discussed in question two, one of the biggest theoretical problems concerns the difficulty of finding a plausible and coherent method of assessing the goodness of outcomes. Let me mention several others here.

One problem concerns the proper scope of moral values. Most agree that at a fundamental level morality requires treating all people equally or impartially. Many think that this implies being neutral regarding space and time. So, for example, there is moral reason to ease *any* innocent person's suffering, *wherever*, and *whenever* it occurs. But many find such neutrality deeply implausible for certain moral ideals. For example, many admit that it is unfair for a poor American, or poor Ethiopian, to be worse off than a rich American who is no more deserving; but they don't think it is unfair for a rich American to be better or worse off than some distant ancestor or future descendent, or someone living now in another galaxy. The problem is to restrict the scope of certain moral ideals across space or time in a plausible coherent way that is compatible with the moral requirement of impartiality.

A similar worry applies regarding different species. If, for example, it really *is* bad, because unfair, for one *person* to be worse off than another no more deserving, why isn't it also bad, because unfair, for one *animal* to be worse off than another no more deserving? But presumably, few, if any animals, *deserve* to be at any level, hence virtually all differences among animals will be undeserved. This raises the question of whether there is a principled way of restricting fairness to rational beings? If not, fairness may have deeply counterintuitive implications.

There are also deep worries about pluralism, which gives weight to a host of different ethical concerns. How does one determine the relative weights of such diverse considerations as freedom, utility, justice, equality, autonomy, rights, and perfection? What about the special obligations we have to people we stand in special relations to, versus the general obligations that we have to all people who share a common humanity, or rationality? How does one trade-off between the categories of care, virtue, consequences, and deontology? These questions are deeply complicated by the fact that different factors may have different weights in different circumstances. So, for example, the (dis)value of suffering may vary depending on whether it is deserved. But what determines how the multitude of different factors matter relative to each other in different circumstances? Without an answer to such questions, how is normative ethics going to guide us?

Additionally, there may be no precise degree to which rights matter relative to perfection, perfection relative to utility, and so on. More generally, moral ideals may only be *roughly* comparable; hence, any theory that takes account of *all* relevant factors may yield woefully incomplete evaluations of different options. Together, these considerations suggest that it may be extremely difficult to arrive at a moral theory that can provide meaningful and determinate direction for most cases.

## Political

Leaders and legislators have powerful political and economic advisors. Some have influential religious advisors. But how many have influential ethics advisors? Similarly, many important ethical problems are addressed by organizations like the United Nations, the World Trade Organization, the World Bank, or the Gates Foundation; but these are largely guided by bureaucrats, and a bevy of political, economic, and scientific advisors. For normative ethics to have a meaningful impact, key decision makers must be informed of, and act on, its results. But this rarely happens.

## Prospects for progress?

Regarding particular applied issues, I am optimistic. Though not widely recognized outside of philosophy, I believe that much progress has already been made on many important applied issues, and that continued work on the problems will almost certainly result in further, substantial, progress. Regarding the theoretical issues, I am less optimistic. Moral theory is extremely difficult, and its deepest problems may prove intractable. Still, theoretical advances in the past 40 years give reason to hope that even those problems that now appear intractable may eventually be resolved. Regarding the political problem, I'm not sure what to think. Many people clamor for decision makers to be guided by ethical considerations. But, unfortunately, much of this clamoring conflates ethics with religion, and too often the zeal for ethical behavior evaporates when acting ethically conflicts with self-interest. Perhaps normative ethicists can help bring about the day where ethics will be taken seriously; but perhaps a sea change in attitudes must first occur before ethicists will be heeded. In that case, one must hope the change occurs soon; for as even my brief list of important applied issues implies, no less than the suffering and premature deaths of billions of beings hang in the balance.[11]

---

[11]I am grateful to Derek Parfit, Jacob Ross, Adam Swenson, Margaret Temkin, and Mike Valdman for many helpful suggestions on earlier drafts of this interview.

# 18

# Peter Vallentyne

## Florence G. Kline Professor of Philosophy
University of Missouri-Columbia, USA

---

**Why were you initially drawn to normative ethics?**

I came late to philosophy and even later to normative ethics. When I started my undergraduate studies at the University of Toronto in 1970, I was interested in mathematics and languages. I soon discovered, however, that my mathematical talents were rather meager compared to the truly talented. I therefore decided to study actuarial science (the applied mathematics of risk assessment for insurance and pension plans) rather than abstract math. After two years, however, I dropped out of university, went to work for a life insurance company, and started studying on my own for the ten professional actuarial exams. When not studying, I would often go to the public library and I was drawn to the philosophy section— although I had no idea of what philosophy was about. I there saw *Logical Positivism*, edited by A.J. Ayer. I was interested in logical thinking and I also favored an optimistic attitude towards life (!) and so I thought that the book might be interesting. I checked it out and was absolutely enthralled with the writings of Bertrand Russell, Rudolf Carnap, Carl Hempel and others (if I'm remembering correctly). Of course, I didn't really understand much of what they were doing, but I did see that they were addressing important problems in a systematic and rigorous manner. I liked it!

I then went on to read most of Bertrand Russell's books and realized that I had a deep interest in philosophy. Eventually, I returned to university—this time at McGill University in Montreal— and completed an undergraduate degree in mathematics and philosophy. After a fantastic year traveling around Greece, I went to the University of Pittsburgh intending to study philosophy of language, philosophy of logic, or philosophy of science. The following year, however, David Gauthier arrived at Pitt and I was exposed to his work and that of John Rawls, Amartya Sen, and John Harsanyi. I was very excited by the applications of the theories of rational and social choice to the foundations of moral theory and I decided to focus on moral philosophy. Initially, my focus was on ethical theory (and consequentialism in particular). Later, I became interested in liberty and equality in political philosophy (and left-libertarianism in particular).

## What example(s) from your work (or the work of others) illustrates the role that normative ethics ought to play in moral philosophy?

I don't really know how to answer this question. So, let me first make some general comments on moral methodology and then identify what I take some of my main contributions to normative ethics to be.

I believe that all the main areas of moral philosophy – metaethics, normative ethics, and applied ethics, for example – can be fruitfully investigated prior to resolution of issues in the other areas. Of course, ideally, we'd resolve the metaethical issues before addressing normative ethics (so that we know what we are talking about!) and ideally the latter would be resolved before addressing applied issues (so that we know what the correct moral principles are!). Given, however, that the issues continue to be highly contested, we can't simply wait for the more basic issues to be resolved. We can fruitfully explore the more derivative issues even if we have to revisit them when more progress is made in more basic issues. Indeed, sometimes the tentative resolution of derivative issues sheds light on the more basic issues.

I am also a big believer in the method of reflective equilibrium in the justification of our judgements—both in general and for morality in particular. I believe that, for any domain of investigation, abstract judgement about principles and concrete judgements about specific cases must be mutually informed and mutually supporting. Our abstract judgements must have a relatively good "fit" with (in the sense of endorsing) our concrete judgements—otherwise they are not suitably anchored in the reality of everyday experience. Our concrete judgements, however, should not be taken as given. They are often ill grounded. They may be based on false beliefs or confusions. Or they may have been well adapted to past circumstances but not to current circumstances. Neither abstract judgement nor concrete judgements should be considered immune to revision in light of pressure from the other. Sometimes concrete judgement should be revised in light of abstract judgement and sometimes vice versa. The method of reflective equilibrium leaves open how much "weight" to give abstract judgements versus concrete judgements. All it rules out is giving no weight to either.

Let me now summarize what I take to be my rather modest contributions to normative ethics. While in graduate school, Geoff Sayre-McCord got me interested in moral dilemmas. I went on to write two papers in which I argued that we need to distinguish between obligation dilemmas and prohibition dilemmas. Obligation dilemmas are situations in which each of two distinct feasible actions is obligatory. These are not conceptually possible as long as obligation is understood in the strong sense that entails permissibility. Prohibition dilemmas are situations in which no feasible action is permissible. Although standard axioms of deontic logic rules this out, this, I claim, is not a matter of deontic logic and but rather a substantive normative view. Thus, we should reject the axiom of deontic logic (e.g., $Per(p) \vee Per(\sim p)$) that rules out prohibition dilemmas as a conceptual matter and focus more on the substantive moral issue about whether they are in fact possible given the nature of morality. As far as I can tell, only few people have picked up on my view, but it seems exactly right to me.

My dissertation (directed by David Gauthier and much influenced by Shelly Kagan) was on the teleological/deontological distinction and several papers came from it. I argued that, rather than there being one fundamental distinction, there were several. A theory can be *goal-directed* without the goal being moral goodness (e.g., Maximize the number of cows in Manitoba!). An *axiological* theory is one that bases the moral permissibility of actions on their moral goodness (e.g., Perform the morally best feasible action!). It does not require, however, that the moral goodness of actions be based on the moral goodness of their outcomes. It could, for example, be based on the agent's intentions. A *teleological* theory is one that bases moral permissibility of actions on the moral goodness of their *outcomes*. Such a theory need not be *maximizing*; it might be satisficing. If we make the usual assumption that moral goodness is *agent-neutral*, then moral egoism (Maximize your own prudential good!) is not teleological. There are, moreover, different conceptions of what counts as an outcome. Three main possibilities are (1) the entire world history that results (including the past), (2) the entire future history, and (3) only the avoidable future (which excludes events that happen no matter what one does). If outcomes are taken to be the entire world history, then a teleological theory can be based on a theory of moral goodness that is sensitive to historical considerations such as desert.

In 1991, I commented on a paper by Mark Nelson that raised problems for utilitarianism (and finitely aggregative theories generally) when the future is infinitely long. For example, one might have a choice between producing a total of 2 units of happiness at each time and producing 1 unit of happiness at each time. Given that both produce infinite totals, neither of which is greater than the other, standard versions of utilitarianism say that neither is better than the other. My thinking about this issue eventually led to several papers on the topic. I argued that the judgement that neither action is better than the other is implausible and defended a revised version of utilitarianism that judges 2 at every time as better than 1 at every time in the infinite case (and agrees with the standard view in the finite case). The rough idea is this: One alternative is better than another if and only if there is some time in the future such that, *for all later times*, the consequences of the first alternative *up to that time* are better (e.g., greater total) than those of the second alternative up to that time. This is a somewhat technical problem, but it shows, I believe, some deep tensions between basic moral principles (e.g., impartiality and Pareto efficiency). It turns out that Frank Ramsey discovered this problem in the 1920s and that economists have developed various solutions that are similar to the one I developed (but much more sophisticated!).

Much of my more recent work is on liberty and equality in the theory of justice. Although this is typically thought of as political philosophy, my view is that it is simply part of normative ethics. The term "justice" is used in different ways, but on one standard usage it stands for those moral duties that we *owe to* individuals (i.e., that correspond to their rights). Justice so understood is that part of ethics that is concerned with *interpersonal* duties. It does not address *impersonal* duties (duties owed to no one).

On the topic of equality, I have argued (with many others) that although equality is a very important moral requirement, it is limited in a number of ways. First, there are constraints imposed by certain rights (e.g., of bodily integrity) on the means by which equality may be promoted. Second, the demand for equality does not require that one promote equality as much as possible (relative to the above constraints); it merely requires that one promote equality sufficiently (where an independent account is needed of what sufficiency requires). Third, justice leaves room for individual accountability for choices. Thus, although some form of equality is required, equality of *outcomes* is not (since that leaves no room for holding agents accountable for their choices). Equality of life prospects (e.g., initial opportunities for wellbeing) and equality of brute luck advantage are two main possibilities. (Brute luck effects are effects that are not attributable to one's choices; e.g., being struck by unforeseeable lightening as opposed to losing money on a lottery ticket.) Both base the relevant equality in part on *initial* effective opportunities, but only the latter also includes *later* outcome brute luck (brute luck in how things later turn out). I have argued that justice does not require that the effects of (later) outcome brute luck be equalized. Instead, at the level of policy, it will be included for instrumental reasons when, and only when, it is efficient to do so (e.g., when administrative costs are low and it is effective in overcoming risk aversion to social desirable activities). Fourth, equality is relevant only for choosing among Pareto optimal (or efficient) options. (An option is Pareto optimal if and only if it is not possible to make someone better off without making someone else worse off. This is a weak notion of efficiency, which requires no interpersonal comparisons of wellbeing.) For interpersonal morality, that is, equality is lexically posterior to Pareto efficiency (i.e., is relevant only when Pareto efficiency is achieved) and thus we are never required to level down to equality. Thus, it is always permissible to make one person better off so long as no one else is made worse off, even if this results in inequality of outcome. This is called Paretian egalitarianism.

Finally, the conception of equality that is relevant for the theory of justice is highly sensitive to *sum-total efficiency* (i.e., favoring the greatest total, which is a much stronger notion of efficiency than Pareto efficiency). All measures of equality hold, as does leximin, that benefits to individuals who remain *below the mean*, no matter how small, take absolute priority (with respect to equality) over benefits to individuals *above the mean*, no matter how large. I argue, however, that a plausible conception of equality for the theory of justice will hold, as does utilitarianism, that the distribution of benefits to *individuals who remain below the mean* should be made so as to *maximize the total benefits*. This gives sum-total efficiency a maximal role in the measure of equality (anything stronger would not be a conception of equality). It avoids any requirement to channel resources to worse off individuals when other individuals below the mean would get greater benefits. On this conception of equality, for example, giving each of two below average people a benefit of two units is more equal than giving a single worse off person a benefit of three units.

On the topic of liberty, I've clarified and defended the thesis of *full self-ownership* and defended a version of *left*-libertarianism. Full self-ownership is simply full ownership applied to the case where the owner and the entity owned are identical. Ownership of an entity consists of control rights (liberty rights to use, claim rights that others not use), compensation rights (rights to compensation if the entity is used without one's permission), enforcements rights (rights to use force to stop others from violating one's rights), transfer rights (rights to lend, rent, give, or sell these rights to others), and immunities to loss of these rights under certain conditions. *Full* ownership of an entity consists of the logically strongest set of ownership rights over that entity that is compatible with someone else having those same rights over the rest of the world. There is, it turns out, some significant indeterminacy in the concept of full ownership, since strengthening rights to compensation and enforcement weakens immunities to loss, and vice-versa. Still, there is a significant determinate core to the concept of full self-ownership, and I have defended its plausibility.

Libertarianism is committed to the natural rights of full self-ownership. This, however, leaves entirely open the moral status of the rest of the world. Right-libertarians (such as Nozick) view it as largely up for grabs by whoever gets there first. Left-libertarians, by contrast, hold that natural resources (all the non-agent resources in the world prior to modification by agents; land, water, air, minerals, etc.) belong to all of us in some egalitarian manner. I have (and am still in the process of) defending a version of equal opportunity for wellbeing left-libertarianism. It holds that individuals have the moral power to appropriate unowned natural resources as long as they pay the full competitive value (based on supply and demand) of the rights that they claim and disburse this payment so as to promote equality of effective opportunity for wellbeing. (Michael Otsuka has also developed and defended a similar view.) This views natural resources as resources to be used for the promotion of equality of opportunity, and further holds the duty to pay the competitive value of rights we claim over natural resources is the only non-consensual source of our duty to help others. Such a view, I have argued, adequately captures the roles of liberty, security, equality, accountability, and prosperity in the theory of justice. Furthermore, it is compatible with the justice of significant state activity (significant taxation, enforcement of rights, provision of public goods, promotion of equality) but not with the justice of the state's prohibition of activities that violate no one's rights.

## How do studies within scientific disciplines contribute to the development of normative ethics?

Biological, psychological, and sociological knowledge can certainly help us understand the nature of happiness, egoism, altruism, moral motivation, free will, etc. I'm somewhat skeptical, however, that such knowledge will prove crucial to resolving any fundamental issues of normative ethics. Empirical information certainly informs fundamental normative judgement but it is rarely decisive in any important way.

There is, however, one scientific discipline that has a lot to contribute to normative ethics. This is normative economics and social choice theory in particular. In the early 1950s, Kenneth Arrow (now a Nobel laureate) proved that it is impossible for the moral (or social) ranking of alternatives (e.g., states of affairs) to satisfy several seeming plausible conditions (transitivity, completeness, Pareto efficiency, non-dictatorship, and independence of irrelevant alternatives). This was an amazing result. Since the 1970s, Amartya Sen (another Nobel laureate) and a significant group of normative economists have greatly expanded the power and range of application of social choice theory. They have systematically clarified issues such as ordinal vs. cardinal wellbeing, interpersonal comparability of wellbeing, equality, sufficiency, rights, freedom, and justice in the space of resources rather than wellbeing. I believe that normative ethics has much to learn from such work. Of course, a lot of this work that takes the form of technical theorem-crunching and this is unlikely to be helpful to normative ethics. Nonetheless, the best work in this area is based on simple and intuitively plausible axioms. The axioms always need to be philosophically assessed, but having crisp statements and results greatly enhances the possibilities of seeing the core issues clearly.

## What do you consider the most neglected topics and/or contributions in normative ethics?

I wouldn't say that the following issue has been neglected, but I do think it is still underdeveloped. What exactly is the difference between an impersonal duty (owed to no one) and an interpersonal duty (owed to someone)? This reduces to the question of what it is to *owe* someone a duty. A natural answer is that the person *has a right* that you perform an action that fulfills the duty. What, however, is it for an individual to have a right against someone? Two broad families of theories of rights have been developed. The choice-protecting family holds that rights protect choices and thus that only autonomous agents have rights. The interest-protecting family holds that rights protect interests and thus that even animals can have rights. My own view is that (1) at the conceptual level, we should recognize both kinds of rights as possible, (2) we should also recognize the conceptual possibility of various hybrids as well, and (3) at the normative level, some kind of hybrid theory is the most plausible. We have, I would argue, rights that protect *both* our choices and our interests, with the protection of choices as lexically prior to the protection of interests. For sentient individuals with no autonomy, this is equivalent to an interest-protecting account. For autonomous agents, however, such an account allows that, where neither consent nor dissent is given (e.g., when there is no time to obtain consent, or when the agent is temporarily unconscious), the interests of the right-holder determine whether a right is violated. Consent, however, is sufficient for non-violation, and dissent is sufficient for violation. Obviously, all this is highly controversial. I merely mention it as an example of where further work is needed.

## What are the most important problems in normative ethics and what are the prospects for progress?

Although it is more a question of metaethics, the nature of critical normativity is one of the most basic unresolved question in philosophy in general. This issue is not particular to moral philosophy. It also includes, for example, prudential normativity and epistemic normativity. It raises issues both about the metaphysics of normativity and of the methodology/epistemology thereof. The issue has been explored at great length with respect to moral normativity (moral realism, non-naturalism, non-cognitivism, etc.), and I'm inclined to think that whatever is correct with respect to morality is also correct generally. Although we've made lots of progress in understanding what the core issues are and what the main positions might be, there is nothing close to agreement. I'm not very optimistic about our being able to solve this one, but we must proceed, I believe, on the presupposition that it can be solved.

One of the most basic problems in normative ethics proper concerns the criteria for moral standing. Do fetuses have moral standing? Do humans that have no potential for experiences (e.g., anencephalic babies or other upper brain dead individuals)? Do possible future people? Do non-human animals? My (very foggy) tentative view is (very roughly) that moral standing is action-relative (a very non-standard view!) and that individuals have moral standing relative to an action if and only if they have some potential for wellbeing (now or in the future) if the action is performed. Upper brain dead individuals have no moral standing relative to any action. Individuals (including non-human animals and sentient fetuses) that already have the capacity (as opposed to potential) for wellbeing have moral standing no matter which action is performed. Presentient fetuses do not have moral standing relative to an action in which they are aborted prior to sentience (and thus are not wronged by such action). A presentient fetus does, however, have moral standing relative to an action for which there is some chance that it will acquire the capacity for wellbeing. This is all messy and complex and I'm here only hinting at a possible position. The main point is that the issue of moral standing is absolutely central to normative ethics. Although it's a difficult issue, I believe that genuine progress is possible. Let's hope!

## Bibliography

Luc Lauwers and Peter Vallentyne "Infinite Utilitarianism: More Is Always Better", *Economics and Philosophy* 20 (2004): 307–330.

Bertil Tungodden and Peter Vallentyne "On the Possibility of Paretian Egalitarianism", *Journal of Philosophy* 102 (2005): 126–54.

Bertil Tungodden and Peter Vallentyne, "Paretian Egalitarianism with Variable Population Size", in *Intergenerational Equity and Sustainability*, edited by John Roemer and Kotaro Suzumura, (Palgrave Publishers Ltd., 2006), ch. 11.

Peter Vallentyne, "Brute Luck, Option Luck, and Equality of Initial Opportunities," *Ethics* 112 (2002): 529–557.

Peter Vallentyne, "Equality, Efficiency, and Priority for the Worse Off", *Economics and Philosophy* 16 (2000): 1–19. Reprinted in *The Economics of Poverty and Inequality*, vol. 1 (International Library of Critical Writings in Economics series), edited by Frank A. Cowell (Cheltenham: Edward Elgar Publishing Ltd., 2003), pp. 112–130.

Peter Vallentyne, "Utilitarianism and Infinite Utility," *Australasian Journal of Philosophy* 71 (1993): 212–7.

Peter Vallentyne, "Two Types of Moral Dilemmas," *Erkenntnis* 30 (1989): 301–318.

Peter Vallentyne, "Teleology, Consequentialism, and the Past," *Journal of Value Inquiry* 22 (1988): 89–101.

Peter Vallentyne, "The Teleological/Deontological Distinction," *Journal of Value Inquiry* 21 (1987): 21–32.

Peter Vallentyne, "Prohibition Dilemmas and Deontic Logic," *Logique et Analyse* 18 (1987): 113–22.

Peter Vallentyne, "Utilitarianism and the Outcomes of Actions", *The Pacific Philosophical Quarterly* 68 (1987): 57–70.

Peter Vallentyne and Shelly Kagan, "Infinite Utility and Finitely Additive Value Theory", *Journal of Philosophy* 94 (1997): 5–26.

Peter Vallentyne and Hillel Steiner, eds., *The Origins of Left Libertarianism: An Anthology of Historical Writings*, New York: Palgrave Publishers Ltd., 2000.

Peter Vallentyne and Hillel Steiner, eds., *Left Libertarianism and Its Critics: The Contemporary Debate*, New York: Palgrave Publishers Ltd., 2000.

Peter Vallentyne, Hillel Steiner, and Michael Otsuka, "Why Left-Libertarianism Isn't Incoherent, Indeterminate, or Irrelevant: A Reply to Fried", *Philosophy and Public Affairs* 33 (2005): 201–15.

# About the Editor

**Thomas S. Petersen**, Assistant Professor of Practical Philosophy, Roskilde University, Denmark. Has published in philosophical journals such as: *Bioethics, Ethical Theory and Moral Practice, Journal of Medical Ethics, Philosophia* and *Theoria.* Co-editor with J. Ryberg and C. Wolf on *New Ways in Applied Ethics,* Palgrave McMillan Publishers, 2007.

**Jesper Ryberg**, Professor of Practical Philosophy, Roskilde University, Denmark. Recent books: *The Ethics of Proportionate Punishment* (Kluwer Academic Publishers, 2004) and *The Repugnant Conclusion. Essays on Population Ethics* (ed.) (Kluwer Academic Publishers, 2005). Has published in philosophical journals such as: *Philosphical Papers, Theoria, Philosophical Quarterly, Res Publica, Ethical Theory and Moral Practice, Journal of Applied Philosophy.*

# Index